Advanced Microsystems for Automotive Applications 2008

Jürgen Valldorf · Wolfgang Gessner (Eds.)

Advanced Microsystems for Automotive Applications 2008

Springer

Dr. Jürgen Valldorf
VDI/VDE Innovation + Technik GmbH
Steinplatz 1
10623 Berlin
Germany
valldorf@vdivde-it.de

Wolfgang Gessner
VDI/VDE Innovation + Technik GmbH
Steinplatz 1
10623 Berlin
Germany
gessner@vdivde-it.de

ISBN 978-3-642-09683-9 e-ISBN 978-3-540-77980-3

DOI 10.1007/978-3-540-77980-3

Cover design: deblik, Berlin

Printed on acid-free paper

9 8 7 6 5 4 3 2 1

springer.com

Preface

With the total number of vehicles steadily increasing and soon approaching one billion, the world is facing serious challenges in terms of both safety of road transport and sustainability. Consequently the two major persistent issues for the automotive industry are improved safety and reduced emissions.

The estimated number of road fatalities is about one million per year. Fast growth of mobility in the developing world and an accelerated urbanisation pose high demands to the automotive industry. Thanks to smart systems anticipating dangerous traffic situations road fatalities will have dropped by more than 30% from 2001 to 2010.

Beyond intensive stock-rearing – with 30% the major contributor to climate change – road traffic is one of the main sectors contributing to climate change: exhaust gases from vehicle engines account for about 20% of the greenhouse gas emissions. Car industry is bearing this challenge and enormous progress has been achieved particularly during the last decade.

The integration of complex microsystems with enhanced intelligence, so-called smart systems has enabled an increase of efficiency of the previously 'dumb' internal combustion engine by on average 1% annually during the last 20 years. In the future, smart systems may help to leverage novel powertrain concepts towards the zero emissions vehicle. Particularly for electric vehicles, anticipatory power management and efficient driving assistance by smart systems will be needed to overcome the range limitations. Adaptive systems predicting the energy demand for a chosen route will help the driver to increase energy efficiency. Networked devices for car-to-car communication could help to avoid the waste of energy due to unnecessary braking and accelerating. Intelligent systems for tire monitoring and control deserve special attention since insufficient tire pressure accounts for more than 3% of the efficiency losses in the car. Future electrical in-wheel motors require new miniaturised functionalities enabled by advanced micro- and smart systems.

Bearing in mind the particular needs of applications for the automobile, stakeholders from the European Technology Platform on Smart Systems Integration (EPoSS) recently compiled a list of future strategic research priorities for next-generation smart systems. These include: advanced functionalities, energy autonomy and networked operation, as well as matters of integration and reliability in harsh environments. The tire monitoring system that comprises an energy harvesting function for autonomous operation presented at the

AMAA 2008 is exemplary for these developments of advanced smart systems devices.

The conference book in hand is a showroom of activities, the AMAA has been known for during the last 12 years: advanced sensors even including one based on the giant magneto resistance (GMR) effect – a finding for which two European physicists were awarded the Nobel prize, recently – several camera and radar systems making road traffic safer by assisting the driver in recognizing pedestrians and obstacles, and human-machine interfaces based on the recognition of hand gestures - an striking example of how smart systems will further enhance the usability of vehicles and the comfort of driving.

I like to express my gratitude to all authors and speakers for their interesting contributions to the AMAA 2008. Financial support was given by the European Commission through their Innovation Relay Center (IRC) Northern Germany which is as gratefully acknowledged as the funding we received from major players in the automotive sector. Kudos has to be given to the members of the Honorary and the Steering Committees for their valuable advice.

And, last but not least, I have to address my particular thanks to Michael Strietzel and Dr. Gereon Meyer of VDI/VDE-IT for their outstanding engagement, and especially Dr. Jürgen Valldorf, the AMAA Chairman, for the organisation of both the event and the publication in hand.

Berlin, March 2008

Wolfgang Gessner

Public Financers

Berlin Senate for Economics and Technology

European Commision

Ministry for Economics Brandenburg

Supporting Organisations

Investitionsbank Berlin (IBB)

mst*news*

ZVEI - Zentralverband Elektrotechnik- und Elektronikindustrie e.V.

Hanser automotive electronic systems

Micronews - The Yole Developpement Newsletter

enablingMNT

MST BW - Mikrosystemtechnik Baden-Württemberg e.V.

Co-Organisators

European Council for Automotive R&D (EUCAR)

European Association of Automotive Suppliers (CLEPA)

Advanced driver assistance systems in Europe (ADASE)

Honorary Commitee

Eugenio Razelli	President and CEO Magneti Marelli S.P.A., Italy
Günter Hertel	Vice President Research and Technology DaimlerChrysler AG, Germany
Rémi Kaiser	Director Technology and Quality Delphi Automotive Systems Europe, France
Nevio di Giusto	President and CEO FIAT, Italy
Karl-Thomas Neumann	CEO, Member of the Executive Board Continental Automotive Systems, Germany

Steering Commitee

Dr. Giancarlo Alessandretti	Centro Ricerche FIAT, Orbassano, Italy
Mike Babala	TRW Automotive, Livonia MI, USA
Serge Boverie	Siemens VDO Automotive, Toulouse, France
Geoff Callow	Technical & Engineering Consulting, London, UK
Bernhard Fuchsbauer	Audi AG, Ingolstadt, Germany
Kay Fürstenberg	IBEO GmbH, Hamburg, Germany
Wolfgang Gessner	VDI/VDE-IT, Berlin, Germany
Roger Grace	Roger Grace Associates, Naples FL, USA
Dr. Klaus Gresser	BMW Forschung und Technik GmbH, Munich, Germany
Henrik Jakobsen	SensoNor A.S., Horten, Norway
Horst Kornemann	Continental Automotive Systems, Frankfurt am Main, Germany
Hannu Laatikainen	VTI Technologies Oy, Vantaa, Finland
Dr. Torsten Mehlhorn	Investitionsbank Berlin, Berlin, Germany
Dr. Roland Müller-Fiedler	Robert Bosch GmbH, Stuttgart, Germany
Paul Mulvanny	QinetiQ Ltd., Farnborough, UK
Dr. Andy Noble	Ricardo Consulting Engineers Ltd., Shoreham-by-Sea, UK
Dr. Ulf Palmquist	EUCAR, Brussels, Belgium
David B. Rich	Delphi Delco Electronics Systems, Kokomo IN, USA
Dr. Detlef E. Ricken	Delphi Delco Electronics Europe GmbH, Rüsselsheim, Germany
Jean-Paul Rouet	Sagem SA, Cergy Pontoise, France
Christian Rousseau	Renault SA, Guyancourt, France
Patric Salomon	4M2C, Berlin, Germany
Ernst Schmidt	BMW AG, Munich, Germany
John P. Schuster	Continental Automotive Systems, Deer Park IL, USA
Dr. Florian Solzbacher	University of Utah, Salt Lake City UT, USA
Bob Sulouff	Analog Devices Inc., Cambridge MA, USA
Berthold Ulmer	DaimlerChrysler AG, Brussels, Belgium
Egon Vetter	Ceramet Technologies Ltd., Melbourne, Australia
Hans-Christian von der Wense	Freescale GmbH, Munich, Germany

Conference chair:

Dr. Jürgen Valldorf	VDI/VDE-IT, Berlin, Germany

Table of Contents

Powertrain

Comfort and HMI

Networked Vehicle

Components and Generic Sensor Technologies

Appendices

Market

Advanced Safety Systems – The Opportunity for Semiconductor Vendors

C. P. Webber, H. Gustafsson, Strategy Analytics

Abstract

Advanced Safety Systems, such as Autonomous Cruise Control (ACC) and Lane Departure Warning (LDWS) have been available for a number of years now. Until recently, fitment has largely been confined to expensive, luxury vehicles. Such systems are now becoming more widely available on mass-market vehicles, such as the Ford Mondeo. These sophisticated electronically controlled systems provide important growth opportunities for semiconductor vendors in maturing regional markets. This paper examines in detail the significant of advanced safety features for automotive semiconductor market. It is based upon extensive vendor research undertaken by Strategy Analytics' Automotive Electronics Service.

1 The Growth in Safety Electronics

For nearly 20 years, Strategy Analytics has been conducting industry-based research into the market for automotive electronics. Although initial electronics applications centred around the vehicle's powertrain, in recent years safety applications have emerged as the fastest growing area of electronics use in the vehicle. Strategy Analytics expects the volume demand for ECUs used in safety applications to grow at 12% per annum over the period 2006 to 2011 [1].

Much of this growth is coming from features that are mandated, such as tire pressure monitoring systems in the USA – but a group of applications known as Advanced Driver Assistance Systems (ADAS) is a significant contributor to growth rates. These systems typically make use of long range sensors such as cameras, radars and lidars to examine the environment around the vehicle, warn the driver of any potentially situations, and ultimately,, take control of the vehicle to either avoid or mitigate an accident.

Examples of ADAS technologies include lane departure warning, autonomous cruise control with collision mitigation/avoidance, blindspot monitoring, night vision, traffic sign recognition and car-to-car communication. Also, ultimately the successful deployment of European E-Call will rely upon it being positioned in the market as a safety system within ADAS.

2 The Importance of New Features to OEMs

Essentially, vehicle manufacturers and their suppliers exist to make a profitable return for their owners and shareholders. There are many factors influencing profitability for an automaker, but one way to improve margins is to sell more high trim vehicles, and more optional extras. An example of the importance of this has been demonstrated recently by Fiat [2], which has managed to change the product mix of its Punto model significantly over the period 2004 to 2006. Top trim level Grande Punto models have around 5.5 times the margin of base models, and by offering the right blend of features, Fiat has managed to raise the proportion of Punto buyers purchasing these models from 5% to 18%.

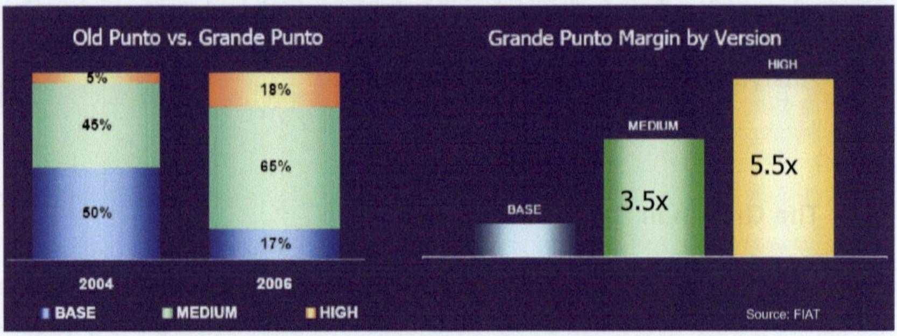

Fig. 1. Higher Trim Models Have Significantly Greater Margins

At present, ADAS technologies are almost exclusively offered as optional features, with price points typically in the €1000 to €2000 range. They thus offer the possibility for automakers to support their margins on the platforms on which they are offered. Profitability for suppliers does not directly flow, as competition to supply these systems is intense, and pressure on pricing levels is high. However, Strategy Analytics' experience over the years indicates that carmakers that are highly profitable make better long-term partners for the supply chain!

3 What Is the Significance for the Automotive Semiconductor Market?

The primary growth drivers for automotive semiconductor demand are positive: Light vehicle production is expected to grow at a CAAGR (Compound Average Annual Growth Rates) of 3.8% over the period 2006 to 2011; automotive electronic system volumes are expected to increase at 8.7% CAAGR over the same period.

Subsequently, the total automotive semiconductor market is forecast to grow by 8% CAAGR from $17.7B in 2006 to $26B by 2011, and reach $31B by 2014 [3], including semiconductor-based sensors (Fig. 2).

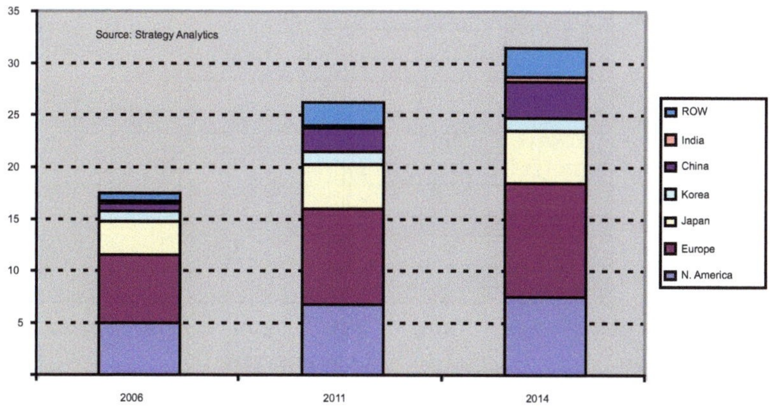

Fig. 2. Automotive Semiconductor Global Market Growth

The overriding picture of long term global light vehicle production is a shift East – both in terms of where the vehicle are being made and in who is making them. This shift East should be of concern to semiconductor vendors primarily dependent upon European and US-based Tier 1 and Tier 2 suppliers that have traditionally had only a small part of their business with Asian-owned brands.

- In 2007, US-owned vehicle manufacturers will account for 28% of global output, European-owned 31% and Asian-owned 39%. By 2014, the US-owned share of global production is expected to have shrunk to 24%, European held steady and 31% and the Asian share climbed to 42%.

The slowing of overall semiconductor market annual growth rates is more pronounced in the most mature vehicle production regions, NAFTA, Europe and Japan. However these regions are also where advanced electronic sensing and control applications will first be deployed by vehicle OEMs. Hence Strategy Analytics analysis has identified these regions as providing the most significant growth opportunities for new semiconductor business demand from existing and emerging safety applications, including ADAS technologies (Fig. 3).

▶ Strategy Analytics' industry research shows a high degree of industry optimism about the success of ADAS features, with fitment levels of systems such as lane departure warning, autonomous cruise control and blindspot monitoring forecast to grow rapidly from their current low base.

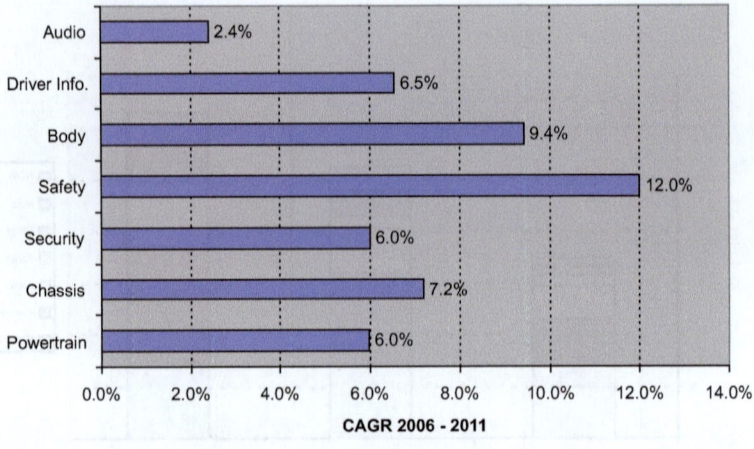

Fig. 3. Automotive Semiconductor Growth by Application

Airbags are the predominant "passive" electronic safety system. Growth from such systems will come from increased deployment in emerging regions, such as Russia, China and India.

ADAS systems are intended to actively enhance vehicle and occupant safety by monitoring the vehicle's immediate environment and to provide driver assistance and communications connectivity. Looking forward, these active systems will provide the highest semiconductor growth opportunity, accounting for 17% of the $1.8 billion global automotive safety application semiconductor demand in 2006, but increasing to 48% of the $4.6 billion market demand by 2014 (Fig. 4).

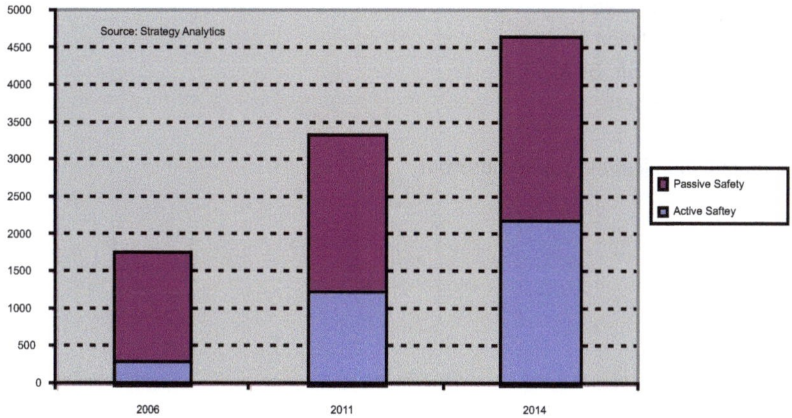

Fig. 4. Automotive Semiconductor Demand from Active and Passive Safety Systems

6 Conclusions

Strategy Analytics' automotive electronics market research has identified safety technologies as the fastest growing application area for automotive electronics. Advanced Driver Assistance Systems play a key part in driving this growth. With overall automotive semiconductor annual demand growth slowing in mature vehicle production regions, ADAS applications provide the most significant growth opportunities for semiconductor IC and sensor vendors.

References

[1] Automotive System Demand 2005 – 2014, Report, Strategy Analytics, Oct 2007
[2] A Fiat brand Progress Report, PowerPoint File, Fiat, Sep 2006
[3] Automotive Semiconductor Demand 2005 – 2014, Report, Strategy Analytics, Oct 2007

Chris Webber
Strategy Analytics
Milton Kaynes, UK
cwebber@Strategyanalytics.com

Hakan Gustafsson
Strategy Analytics GmbH
Landsberger Str. 394
81241 München
Germany
hgustafsson@strategyanalytics.com

Safety

Dual Frequency Methods for Identifying Hidden Targets in Road Traffic

A. Fackelmeier, C. Morhart, E.M. Biebl, Technical University Munich

Abstract

Protection of vulnerable road users (VRU) is a key issue of future driver assistance systems. There, an important point is to determine whether a VRU is visually hidden to the driver. This additional information is essential for the driver to prevent an accident or to reduce its severity. In this paper we show that by using a transponder attached at the VRU, wavelength dependent diffraction effects can be utilized to distinguish between visible and hidden VRUs. Several diffraction measurements were carried out taking place in open space areas with simple big scattering objects and in different road traffic scenarios demonstrating a reliability of nearly 100%. If future driver assistance systems for detecting VRUs are complemented with automotive electronics which enable distinction between visible and hidden VRUs, accidents can be avoided [5]

1 Introduction

Protection of vulnerable road users (VRU) is a key issue of future driver assistance systems. Currently, several different systems are under investigation, based on passive sensing techniques (radar, lidar, camera) as well as on cooperative sensor systems. In a cooperative system the VRU is required to wear a transponder in order to identify themselves as a VRU as well as to send a beacon signal for localization. In contrast to passive sensing, the beacon signal can also be detected if the person wearing the transponder is visually hidden, for instance by a parked car or a roadside tree. Of course, information about this situation, i.e. whether the transponder is visually hidden or not, will be of great value for alerting the driver of the car. If the person wearing the transponder is visually hidden, the situation is much more hazardous.

Our method to identify visually hidden persons is based on dual-frequency discrimination. A transponder (according to the terms of radar technology we call it "active target") attached at the VRU emits two signals with different frequencies which are detected by the receiver in the car. If the VRU is invisible for the driver, the signals are diffracted around the scattering object and get an additional attenuation. Signals with different wavelengths experience different attenuation at different observation angles in the shadow zone due to the wavelength dependent diffraction. The difference in signal power at the two frequencies compared to the difference at line of sight (LOS) received at the driver side is an indicator whether the emitter is hidden or not. There are many parameters to be investigated affecting the measurement result such as orientation of the VRU to the scattering object and to the receiver, height of the antennas, multipath propagation due to buildings or objects nearby and the size and shape of the scattering object which hides the VRU. Our objective was to determine whether this diffraction effect is measurable in real road traffic scenarios, how reliable a detection of hidden VRU is, and how the parameters mentioned above influence the measurement results.

2 Wave Propagation Effects

For theoretical analysis, the dimension of this scattering problem compared to the wavelength is too big for applying EM field simulation methods. The Geometrical Theory of Diffraction (GTD), first introduced by Keller in 1962 [3], is an extension to the geometrical optics which adds diffracted rays E_d to the geometrical optics [1].

$$E_d = R(\gamma, \varphi_1, \varphi_2, \lambda, \text{polarization}) \cdot A(s_d) \cdot E_i \cdot e^{-jk_0 s_d} \tag{1}$$

Dimensional considerations show that the edge diffraction is the strongest besides corner and surface diffraction [3] and therefore, the latter are neglected. Edge diffraction occurs with an edge radius small compared to λ. The incident wave is diffracted at the edge and spreads out as a cylindrical wave.

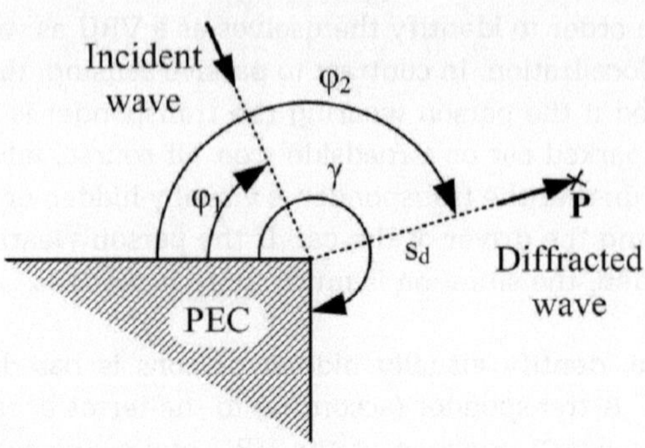

Fig. 1. Edge diffraction

With the diffraction coefficient R, the signal power difference in the shadow region can be calculated for different frequencies. E_i is the field incident at the point of diffraction and $A(s_d)$ (= $1/\sqrt{s}$ for edge diffraction) is the spreading factor with the distance s_d between the scattering point and the observation point P were the diffracted field strength E_d is present. Between f_u = 433 MHz and f_o = 2.45 GHz the ratio in signal power of the diffracted waves is about 7.5 dB.

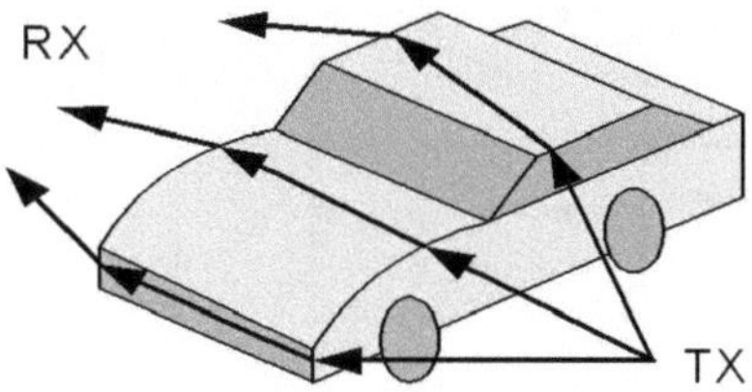

Fig. 2. Double edge diffraction at a car

If there is a second edge, double edge diffraction occurs and the power difference amounts to about 15 dB. Double diffraction occurs quite often in road traffic situations with cars as scattering objects (Fig. 2). The mentioned frequencies were chosen because a bigger difference in frequency results in a bigger difference in signal power of the diffracted wave. Lowering the frequency f_u to the next lower ISM band which ranges at 40.70 MHz brings no significant increment in signal power difference but unfavourably bigger antennas. A higher frequency for f_o can be advantageous but this has not been investigated.

Multipath propagation can cause an attenuation of more than 15 dB which is very problematic in this application as it has to be distinguished between attenuation due to scattering objects in the signal path and attenuation due to multipath propagation. In road traffic situations, multipath propagation can occur through every object near the road such as buildings, vehicles or other obstacles nearby.

3 The RF-Power Measurement System

A RF-power measuring system was developed and built up for the ISM-frequency bands at 2.45 GHz and 433 MHz with commercial IC devices to show that the target detection is realizable also without high-precision measurement equipment. For each emitter (TX), Phased Locked Loop (PLL) frequency stabilization is used to avoid a frequency drift which can result in an output power variation and therefore decrease the measurement accuracy. At the receiver (RX) a high dynamic range is necessary to cover different distances of RX and TX and the additional attenuation due to the scattering objects in between. A low pass filter is implemented to flatten the interfering fast fading peaks in the signal waveform.

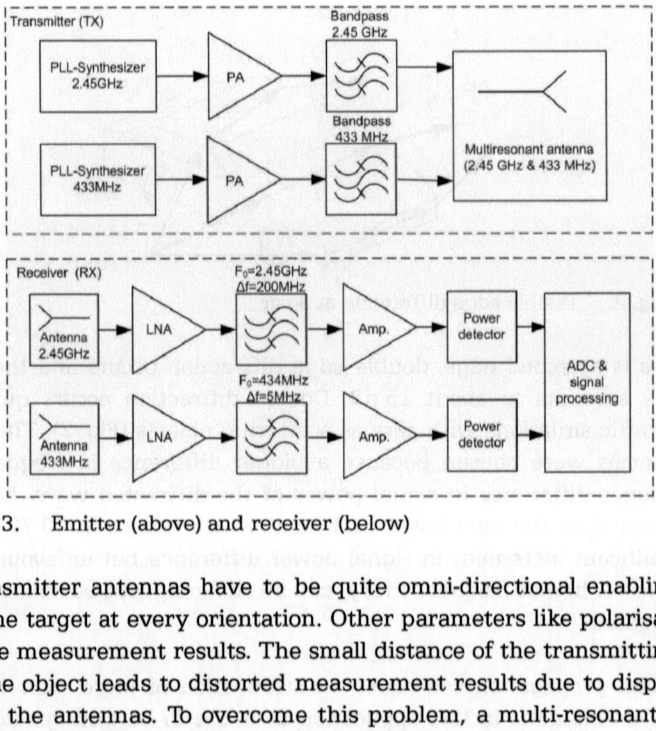

Fig. 3. Emitter (above) and receiver (below)

The transmitter antennas have to be quite omni-directional enabling detection of the target at every orientation. Other parameters like polarisation also affect the measurement results. The small distance of the transmitting antennas to the object leads to distorted measurement results due to displacement between the antennas. To overcome this problem, a multi-resonant antenna was developed which provides a nearly coinciding phase centre.

4 Detection in Case of Single Objects in the Signal Path

To verify radiation effects, several measurements with simple objects were carried out in open space areas. The transmitter (TX) emits continuous wave signals at both frequencies that travel to the receiver (RX) either directly or disturbed by an scattering object. A relative movement of the target, receiver or object causes a position-dependent signal power at the receiver antennas. This signal power is measured separately for both frequencies at RX. The scattering object used is a metal sculpture with sharp edges and dimensions of 2 m in each direction.

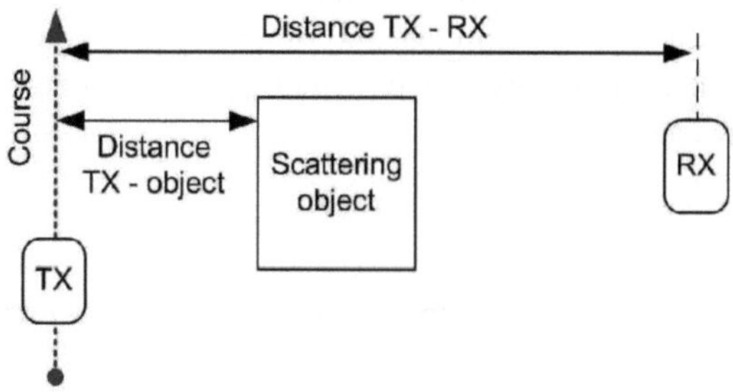

Fig. 4. Measuring setup for detection of a target hidden by a single object

Fig. 5 shows the diffraction measurement with a distance of 35 m between the receiver and the target and 1 m distance between target and object. After measurement point MP 200, the target person gets invisible for the receiver, and there are only waves present at RX which are double diffracted at the edges of the scattering object. The 433 MHz signal experiences only a small attenuation due to the object compared to the 2.45 GHz one. The lower graph indicates the ratio in signal power of both frequencies. As the theory predicts, this value is about 15 dB if the target is hidden relating to 0 dB for line of sight (LOS). If the threshold is placed for instance at -5 dB, it is easy to figure out in which interval the object is hidden by the target.

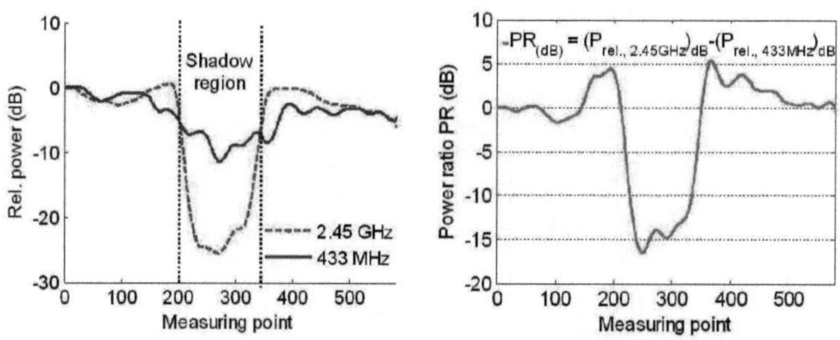

Fig. 5. Detection of a target hidden by a single object above: normalized received power at the two frequencies, below: power ratio for the detection ($distance_{TX\text{-}RX}$ = 36 m, antenna $hight_{TX}$ = 1.2 m, antenna $hight_{RX}$ = 0.8 m, ground: mown grassland)

At certain distances, strong peaks in signal power arise in the shadow region resulting from interference between waves diffracted around different sides of the scattering object which can be verified through comparing propagation paths. At scattering objects with big radii, creeping waves have been detected. This has been verified through calculation of ray propagation paths leading to points with destructive interference which agree with measured peaks in the diffracted wave signal [4].

5 Detection of Hidden Targets in Road Traffic Scenarios

Several different road traffic scenarios have been analyzed to verify that the results are repeatable. Figure 6 shows a typical situation - parked cars at the roadside. In the first measurement, the emitter (TX) moves normal to the road in a certain distance to the parking car. The receiver (RX) is placed at a fixed position at the road. With this configuration, the scenario of a VRU moving from invisible positions to visibility is analyzed at a fixed position of the car. The measurement starts at d_1 = 10 m corresponding to the measuring point (MP) zero in figure 7, where the target is hidden by the cars. After MP 390, the target person becomes visible for the receiver, and TX and RX are in LOS. By setting the threshold again at -5 dB, the target person can be identified as hidden (before MP 390) or as visible (afterwards). In this case, the detection has a reliability of about 95% for the target positions regarded. By averaging the signal over a certain time period, the error probability can be minimized.

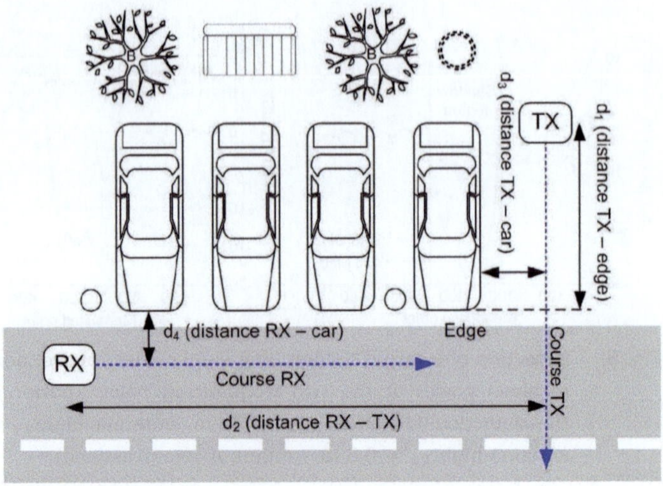

Fig. 6. Measurement configuration for traffic situations

Disturbing ripples of up to 20 dB arise due to constructive and destructive interference between direct and reflected waves in LOS case between diffracted and reflected waves or also between several diffracted waves in the shadow region. By averaging the signal over a certain period of time, the power ratio between the two frequencies for hidden targets amounts always to about 15 dB relating to 0 dB for LOS. This result was confirmed in different measurements with moving targets and coincides with the theoretical calculation of chapter 2.

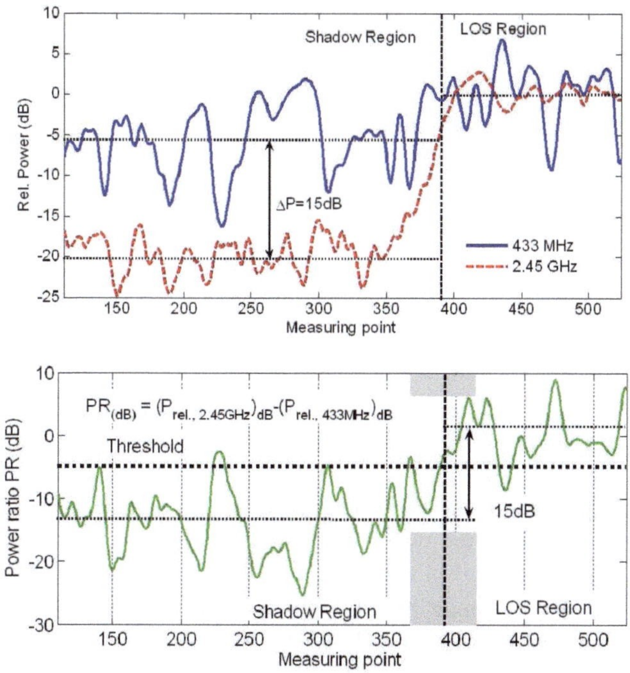

Fig. 7. Detection measurement of a moving target; above: normalized received power at the two frequencies, below: power ratio for the detection (d_2 = 30 m, d_3 = 1 m, d_4 = 1 m, antenna $heigh_{TX}$ = 0.8 m, antenna $height_{RX}$ = 0.6 m, ground: asphalt)

The following measurement takes place in the same scenario indicated in figure 6 but with a moving receiver (RX) and an emitter at fixed positions. The receiver moves parallel to the street as can be seen from Fig. 6. During this measurement, the emitter remains at a fixed position. Two measurement results with different target - edge distances (d_1) have been put together in Fig. 8: $d_{1,LOS}$ = 0 m (target just visible) and $d_{1,HIDDEN}$ = 2 m (target hidden). With this configuration the scenario of a moving car is analyzed in two cases – one with a hidden VRU and one with a VRU just visible.

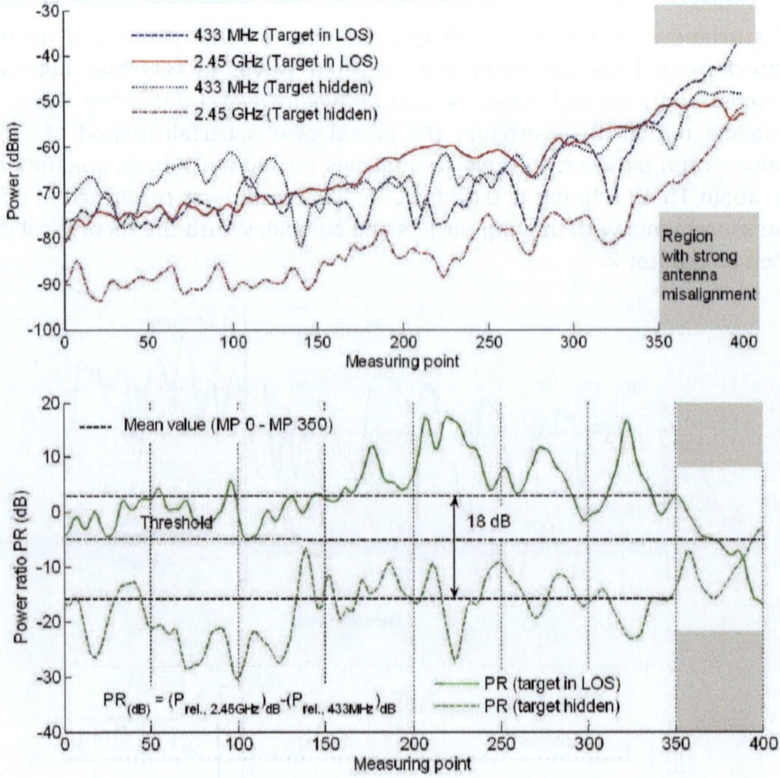

Fig. 8. Detection measurement of a moving target, above: received power at the two frequencies, below: power ratio for the detection (d_3 = 1 m and d_4 = 1 m, $d_{1,LOS}$ = 0 m (target just visible) and $d_{1,HIDDEN}$ = 2 m (target hidden) antenna height$_{TX}$ = 0.8 m, antenna height$_{RX}$ = 0.6 m, ground: asphalt)

The movement starts at a target distance of d_2 = 35 m (MP 0) and stops at d_2 = 2 m (MP 400). At the last few meters there is a big misalignment of the antennas because TX is displaced from the RX moving path by 3 m (target hidden) or 1 m (target in LOS) respectively. So these values after MP 350 are not used for calculation of the mean value. In contrast to the prior measurements which show relative power levels this time the absolute signal power as a function of distance is shown in the upper graph of figure 8. The difference in the mean values of the power ratio in LOS and hidden case range with 18 dB in the same size as in the previous measurements. The signal difference is at all distances big enough for detection of hidden targets. By setting the threshold again to –5 dB, the detection has a reliability of more than 95%.

The height of the antenna over the ground has a big impact on the measurement results. If both antennas are at a height where LOS can be achieved through the windows of the cars, the signal difference is less because the signals experience double slit diffraction. For 2.45 GHz, the slit (car window) is big enough that at RX a signal power level is measured which is even higher than for LOS because of Fresnel diffraction. Otherwise, if the antennas are placed too low above the ground, destructive interference of the direct signal with the signal reflected on the ground occurs and Frii's transmission equation gets invalid and has to be replaced by the two-ray model. In this case, at certain distances, the 2.45 GHz signal is attenuated up to 20 dB, and the detection will be incorrect.

5 Conclusion

In this paper, the detection of hidden persons by using a transponder attached at the person is demonstrated. It has been shown that wavelength dependent diffraction effects can be utilised for the detection whether an emitter is hidden. In order to show that this effect is measurable without measurement equipment, a RF-measuring system was developed and built up for 2.45 GHz and for 433 MHz. In open space areas with the used system configuration, detection of targets hidden by single big objects can be achieved up to 60 m distance between TX and RX. Here the power difference amounts to about 15 dB between hidden emitters relating to 0 dB for LOS confirming the theoretical calculation of chapter 2.

Several target detection measurements in different road traffic scenarios have been carried out. The strong ripple in the power measurement arises from multipath propagation. By averaging the signal over a certain period of time, the power ratio between the two frequencies for hidden targets amounts to about 12 -18 dB relating to 0 dB for LOS.

The detection whether a target is hidden behind a car is possible at any orientation of the car but depends on the height of the emitter. Relatively small objects like persons or street lamp poles in the signal path do not lead to detection which may be desirable.

The results were confirmed in many different measurements with moving targets and also with moving receivers. Multipath propagation causes strong interferences. Especially in situations with buildings or other objects at the roadside, a detection with-non averaged measurement values is in some cases impossible because of the strong ripple in signal power. A solution to overcome this problem is using diversity reception. In this case, for the reception of a continuous wave signal only, this can be accomplished by switching two or more antennas consecutively to one LNA and always taking the highest signal value.

An important aspect is to set the threshold at a level with the best detection reliability between LOS and hidden target. A variable threshold depending on other parameters such as distance between RX and TX which has to be measured simultaneously may be advantageous.

If the body of the target person turns into the signal path, both frequencies are attenuated up to 30 dB. This attenuation does not arise at the same turning angle for both frequencies, so an additional power difference of more than 15 dB must be taken into account, too. For this reason, the error probability in detection for statistical target orientation increases significantly. Further investigations have to be made to determine the impact of the target person's orientation at different road traffic situations.

References

[1] Borovikov, V. A.; Kinber, B. Y.: Geometrical theory of diffraction. Bath: Bookcraft, 1992.

[2] Hecht, E.: Optik. 4th ed. München: Addison-Wesley Publishing Company, 2002.

[3] Keller , J. B.: Geometrical Theory of Diffraction. Journal of the Optical Society of America, volume 5, number 2, pp. 116-132, February, 1962.

[4] McNamara, D. A.; Pistorius, C. W. I.; Malherbe, J. A. G.: The Uniform Geometrical theory of diffraction. Boston: Artech House, 1989.

[5] Verkehr - Kinderunfälle im Strassenverkehr 2005 [online, 5462405057004.pdf]. Copyright: Statistisches Bundesamt, last update: 4. Oct. 2006, available at: https://www-ec.destatis.de.

Andreas Fackelmeier, Christian Morhart, Erwin Biebl
Technische Universität München, Fachgebiet Höchstfrequenztechnik
Arcisstr. 21
80333 München
Germany
fackelmeier@tum.de
morhart@tum.de
biebl@tum.de

Keywords: diffraction, automotive, hidden, active target, ISM Band

Multi Spectral Pedestrian Detection and Localization

G. Bauer, F. Homm, L. Walchshäusl, BMW Group
D. Burschka, Technische Universität München

Abstract

A fusion method of multi spectral video sources for pedestrian localization is presented. The question of how an image from a far-infrared (FIR) camera can be registered with an image from a CMOS video camera to extract distance information is discussed. Because the nature of thermal images is quite different to standard video images, one of the biggest challenges is to find mutual information from each of the cameras that can be combined. A new approach for multi-modal stereo-matching based on contour information as common feature is introduced. In the first step, the object contour is extracted on hot spots in the FIR image by means of extended active contour models. In the second step the stereo correspondence problem is solved with a fast active contour shape matching algorithm utilizing the epipolar constraint. Finally, a postponed image classification based on histograms of gradients decides if the region of interest encloses a relevant object such as a pedestrian.

1 Introduction

Currently, several automotive manufacturers offer night vision systems based on far infrared technology (FIR). These systems support the driver in low-light situations by highlighting non- or sparsely illuminated living objects such as pedestrians and animals. FIR systems use an imager which directly measures the thermal radiation of objects. The primary advantage of a FIR imager is the high-contrast mapping of objects like pedestrians, animals and vehicles which are warm in comparison to the background. This enables both an improved visibility of far distanced objects to the driver and an enhanced recognition potential for an automated image vision system.

Future driver assistance systems will require increased reliability in environmental perception and data representation of the real world. Systems like collision warning or autonomous braking for example also require accurate distance estimates of relevant objects. With a monocular vision system, only a rough longitudinal position information can be estimated by means of model assumptions like the flat world assumption. However, with multi-camera systems an improvement of the classification and localization quality is achievable. Although depth information can be deduced through a disparity analysis of multiple images, it is not desirable to purchase two FIR-cameras, due to the additional costs and limited installation space in the car. Therefore, it would be very attractive to combine data from a FIR camera with a CMOS video camera since there may be a video camera already in the car for other driver assistance systems (e.g. lane departure warning). The major difficulty is that most well-known standard stereo algorithms that calculate disparities based on image intensities would fail.

The paper at hand describes a multi-camera computer vision algorithm that processes images taken with a multi spectral stereo system which is composed of a FIR and a video camera. The nature of FIR-images is quite different from standard vision-based images. On surfaces with homogeneous temperature and thermal emissivity, the resulting image will appear unstructured in the FIR image while the object may have a textured pattern in the visible spectrum. Conversely, an object that is untextured in the visible spectrum but has a distribution in surface temperature would look quite structured in a FIR image. Thus, texture information is not suitable to combine these mixed video-sources.

Consequently, the introduced approach operates solely on contour information. The presented solution utilizes pedestrian hypotheses which are provided by a monocular FIR system. In a further step it extracts the object contour by means of active contour models. A high performance shape matching algorithm based on epipolar constraints calculates a corresponding match of the contour in the video image. Finally, the distance information of the model can be deduced via triangulation.

2 Multi-Spectral Stereo Algorithm

Since texture information is not suitable to combine multi modal video-sources we solely operate on contour information. Active-Contour Models, first introduced by Kass [1], are curves defined within an image domain that can be deformed under the influence of internal and external forces to lock onto features of interest in an image. Usually, the features are object boundaries defined by an edge map. Active Contours are widely used in many applications, including edge detection [1], shape modeling [2], segmentation [3], and motion tracking [4]. Given an initial guess of the pedestrians boundary in an image, active contour models can find the "actual" boundary of an object and with some modifications to the energy formulation they can be used in a fast shape matching algorithm to solve the stereo correspondence problem for multi spectral images.

2.1 Parametric Active-Contour Model

A parametric Active-Contour Model [5] is represented as an ordered collection of n contour points in the image plane:

$$\begin{aligned} \mathbf{V} &= \{\mathbf{v}_1,\ldots,\mathbf{v}_n\} \\ \mathbf{v}_i &= (x_i,y_i), \qquad i=\{1,\ldots,n\} \end{aligned} \tag{1a}$$

where $\mathbf{v}_i$ corresponds to the image point (x_i,y_i). Every point $\mathbf{v}_i$ has one predecessor $\mathbf{v}_{n-1}$ and one successor $\mathbf{v}_{n+1}$, where $\mathbf{v}_0 = \mathbf{v}_n$ and $\mathbf{v}_{n+1} = \mathbf{v}_1$. In the practical use case of representing a pedestrians shape as a deformable contour, a first initial approximation is needed. Due to the large number of variation in appearance, a simplified and general model turns out to be the best. Particularly, a centered ellipse representation is chosen (see Eq. 1(a)).

In a second step the initial active contour model is modified iteratively from its position through a process of local minimisation of an energy functional. For each image point in an rectangular neighborhood of $\mathbf{v}_i$, an energy term is computed:

$$\mathbf{E}_i = \underbrace{\alpha\,\mathbf{E}_{con}(\mathbf{v}_i) + \beta\,\mathbf{E}_{curv}(\mathbf{v}_i)}_{\textit{internal forces}} + \underbrace{\gamma\mathbf{E}_{img}(\mathbf{v}_i)}_{\textit{external forces}} \tag{1b}$$

where the "internal forces" are coming from within the curve itself and the "external forces" are computed from the image data. α, β and γ are constants providing the relative weighting of the energy terms and $\mathbf{E}_i$, $\mathbf{E}_{con}$, $\mathbf{E}_{curv}$ and $\mathbf{E}_{img}$ are matrices $\in \mathbb{R}^{m\times m}$. The value at the center of each matrix corresponds to the contour energy at the image point $\mathbf{v}_i$. Other values in the matrices, named as $\mathbf{E}(\mathbf{v}_i)^{j,k}$, correspond to the energy at each point in the $m \times m$ neighborhood of $\mathbf{v}_i$ while $\mathbf{p}_i^{j,k}$ denotes the image points around $\mathbf{v}_i$. The definitions of $\mathbf{E}_{con}$ and $\mathbf{E}_{curv}$ have been adopted from [6]. The external force $\mathbf{E}_{img}$ is the common formulation of the negative gradient magnitude.

$$\mathbf{E}_{con}(\mathbf{v}_i)^{j,k} = \bar{d} - \|\mathbf{p}_i^{j,k} - \mathbf{v}_{i-1}\|_2 \quad (1c)$$

$$\mathbf{E}_{curv}(\mathbf{v}_i)^{j,k} = \|\mathbf{v}_{i-1} - 2\mathbf{p}_i^{j,k} + \mathbf{v}_{i+1}\|_2 \quad (1d)$$

$$\mathbf{E}_{img}(\mathbf{v}_i)^{j,k} = -|\nabla f(\mathbf{p}_i^{j,k})| \quad (1e)$$

The Eq. (1c) tries to minimize the distance between the contour points which has the effect of causing the contour to shrink. The first term is the average euclidean distance between all points $\mathbf{v}_i$. The Eq. (1d) forces the contour to be smooth and Eq. (1e) attracts the contour toward the strongest image gradient $f(\ldots)$. At each iteration step the energy function $\mathbf{E}_i$ is evaluated for the current point $\mathbf{v}_i$ and for the points $\mathbf{p}_i^{j,k}$ in its $m \times m$ neighborhood. All contour points $\mathbf{v}_i$ are moved to the position in their neighborhood which attains the minimum energy, assuming that this local solution approaches the global optimum. The iteration process is aborted if the length of the model outline $u = \sum_{\forall i} \|\mathbf{v}_i - \mathbf{v}_{i+1}\|_2$ does not change any longer or if a maximum amount of iteration steps is reached (see Fig. 1). These terminating conditions ensure both a stable model shape and an abort in case of incorrect behavior.

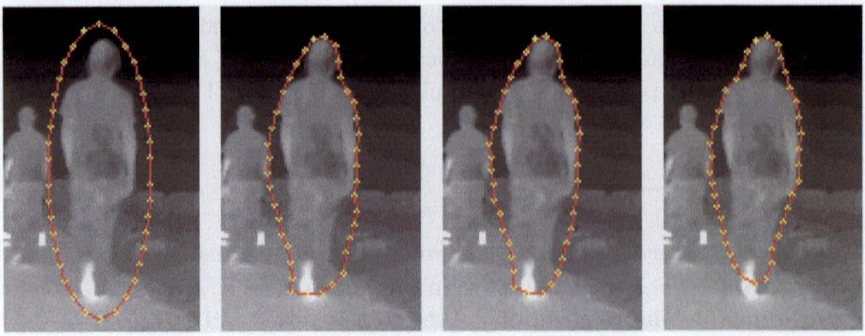

Fig. 1. Approximation of a pedestrian shape with an active contour model after 0 (a, left), 16 (b, middle left), 24 (c, middle right) and finally in the stable stage after 32 (d, right) iterations. The contour points $\mathbf{v}_i$ are marked by yellow crosses.

2.2 Improvements

In the conventional Active-Contour Models as defined in (1b), only the gradient magnitude contributes to the "external forces" energy. However, the simple force definition of E_{img} has some major disadvantages. When an image has a complex background, the active contour gets confused by including strong background gradients into the minimisation process (see Eq. 1(d)). In order to improve the effectiveness of the active contour, it is helpful to use all available information. It is assumed, that a pedestrian in a FIR Image always appears warmer and therefore with higher intensity values than the surrounding background objects. The image gradient direction is defined as

$$\theta_G(\mathbf{v}_i) = \arctan 2\left(\frac{g_y(\mathbf{v}_i)}{g_x(\mathbf{v}_i)}\right) \tag{1f}$$

where $g_x(\mathbf{v}_i)$ is the horizontal and $g_y(\mathbf{v}_i)$ the vertical image gradient at position (x_i, y_i). The gradient map can be obtained by applying an arbitrary gradient operator to the image, such as Sobel or Prewitt.

First, the absolute difference θ_Δ between the image gradient direction $\theta_G(\mathbf{v}_i)$ and the model orientation $\theta_M(\mathbf{v}_i) = \arctan 2\left(\frac{y_{i-1}-y_{i+1}}{x_{i-1}-x_{i+1}}\right)$ is computed by:

$$\theta_\Delta(\mathbf{v}_i) = \begin{cases} (\pi - |\theta_G(\mathbf{v}_i) - \theta_M(\mathbf{v}_i)|) \mod \pi & \theta_\Delta(\mathbf{v}_i) > \pi \\ |\theta_G(\mathbf{v}_i) - \theta_M(\mathbf{v}_i)| & \text{else} \end{cases}$$

Finally, the definition of the external force from Eq. (1e) is replaced by

$$E_{img}(\mathbf{v}_i)^{j,k} = \begin{cases} -|\nabla f(\mathbf{p}_i^{j,k})| & \theta_\Delta(\mathbf{p}_i^{j,k}) \leq \varepsilon \\ 0 & \text{else} \end{cases} \tag{1g}$$

where image gradient magnitudes are excluded whose directions differ more than a threshold ε from the active contour model gradient direction.

This redefinition of the extended force equation improves the result of the contour approximation in many practical cases, but there is still one major problem. Caused by the limited capture range of the $m \times m$ matrix $\mathbf{E}_{img}$ and the local property of the edge magnitude ∇f, active contours have difficulties progressing into boundary concavities. For example, the head-shoulder body part of a pedestrian is such a region. The basic idea is to increase the capture range of the external force field to guide the active contour toward the desired boundary. One solution is to apply a nonlinear transformation to the edge magnitude map. This transformation, known as "distance potential forces" [7] does not change the direction of the gradients, only their magnitudes. A better solution that addresses to that problem is the gradient vector flow (GVF) [8, 9]. The approach extends the gradient map farther away from the edges and into homogeneous regions using a gradient directional preserving diffusion process (see Fig. 2(a) and 2(b)). The gradient vector flow field is the vector field $\mathcal{V}(x,y) = [u(x,y) = g_x(x,y),\, v(x,y) = g_y(x,y)]$ that minimizes the energy functional:

$$\mathscr{E} = \int\int \mu(u_x^2 + u_y^2 + v_x^2 + v - y^2) + |\nabla f|^2 |\mathscr{V} - \nabla f|^2 dxdy. \tag{1h}$$

To obtain the corresponding dynamic active contour equation, the external force $-|\nabla f(x,y)|$ in Eq. (1g) and the gradient direction $\theta_G(x,y)$ in Eq. (1f) are simply replaced by:

$$-|\nabla f(x,y)| = \sqrt{u(x,y)^2 + v(x,y)^2} \tag{1i}$$

$$\theta_G(x,y) = \arctan 2\left(\frac{u(x,y)}{v(x,y)}\right) \tag{1j}$$

2.3 Active-Contour Matching with Epipolar Constraint

The objective of the shape matching algorithm is to find a set of corresponding points $\mathbf{V}' = \{\mathbf{v}_1' \ldots \mathbf{v}_n'\}$ in the video image that maximizes some cost function. It is assumed, that this set of points is the most feasible match of the extracted active contour shape $\mathbf{V}$ in the FIR image. In the literature, there are many investigations devoted to the problem of matching a shape represented by a ordered set of points to an image [10, 11, 12]. However, the results by whichever method are not satisfactory because the correspondence problem $\mathbf{v}_i \leftrightarrow \mathbf{v}_i'$ is not straightforward to find a fast uniquely one-to-one mapping between the two images. Our stereo matching algorithm solves the correspondence problem by stepwise model shifting in a predefined search window utilizing the epipolar geometry.

Let $\mathbf{X}_i$ be a 3D Point, $\mathbf{P}_1\mathbf{X}_i = \mathbf{x}_i$ and $\mathbf{P}_2\mathbf{X}_i = \mathbf{x}_i'$ be its projections into image 1 and image 2. $\mathbf{P}_1$ and $\mathbf{P}_2$ are two projection matrices $\in \mathbb{R}^{3\times4}$. For every point $\mathbf{x}_i$ its corresponding point $\mathbf{x}_i'$ satisfies the epipolar constraint:

$$\mathbf{x}'^T\mathbf{F}\mathbf{x} = 0 \tag{2a}$$

where $\mathbf{F}$ is known as the fundamental matrix [13]. If the relative camera geometry is known, then, given a point $\mathbf{v}_i$, its epipolar line $l_i' = Fv_i$ can be computed and the search space can be restricted to the line l_i'. In the practical case of searching a pedestrian model, which is a representation of an 3D object where the minimum and maximum dimensions h_{world}^{min}, h_{world}^{max} in the real world are known, its corresponding model point v_i' has only to be searched along a restricted part of l_i' rather than the whole epipolar line (see Fig. 3). If the height h_{world}, the focal length f and the height in the image plane h_{image}, which is the major axis of the model V, is known, these dimensions can be directly transformed to a distance interval $D = [d_{min} \ldots d_{max}]$ where $d = \frac{fh_{world}}{h_{image}}$.

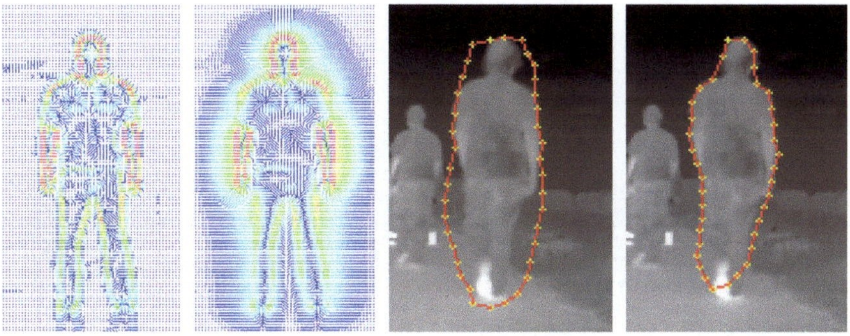

Fig. 2. Gradient vector field of a FIR pedestrian with bidirectional gradient direction and a decreasing gradient strength from red to blue. Gradient vector field with GVF after 0 (a, left) and 120 (d, right) iterations. Approximation of a pedestrian shape with an active contour model based on GVF after 8 (a) and finally in the stable stage after 24 (d) iterations.

The next step is to compute a set of models

$$\mathcal{M}_{search} = \{\underbrace{(\mathbf{v}_1'^1, \ldots, \mathbf{v}_n'^1)}_{\mathbf{v}'^1}, \ldots, \underbrace{(\mathbf{v}_1'^m, \ldots, \mathbf{v}_n'^k)}_{\mathbf{v}'^k}\} \tag{2b}$$

which cover the entire restricted search space in the video image. $\mathcal{M}_{search}$ can be obtained by applying a back-projection to every contour point $\mathbf{v}_i$ in the first image. A more accurate and faster solution can be obtained by using a back-projection only for d_{min} and d_{max}.

$$\begin{aligned} v_i'^{min} &= P_2 X_i(d_{min}) \\ v_i'^{max} &= P_2 X_i(d_{max}) \end{aligned} \tag{2c}$$

For the n model points the distance $\delta_i(x,y) = |v_i'^{max} - v_i'^{min}|$ is multiplied by a scaling factor $\alpha < 1$. By means of the step-size width $\delta_i(x,y) = \alpha\delta_i(x,y)$ all points between $v_i'^{min}$ and $v_i'^{max}$ can be calculated by linear interpolation.

$$\begin{aligned} \mathcal{M}_{search} &= \{\underbrace{(v_1'^{min}, \ldots, v_n'^{min})}_{V'^{min}}, \\ &\underbrace{(v_1'^{min} + \delta_1, \ldots, v_n'^{min} + \delta_n)}_{V'^2}, \ldots, \underbrace{(v_1'^{max}, \ldots, v_n'^{max})}_{V'^{max}}\} \end{aligned} \tag{2d}$$

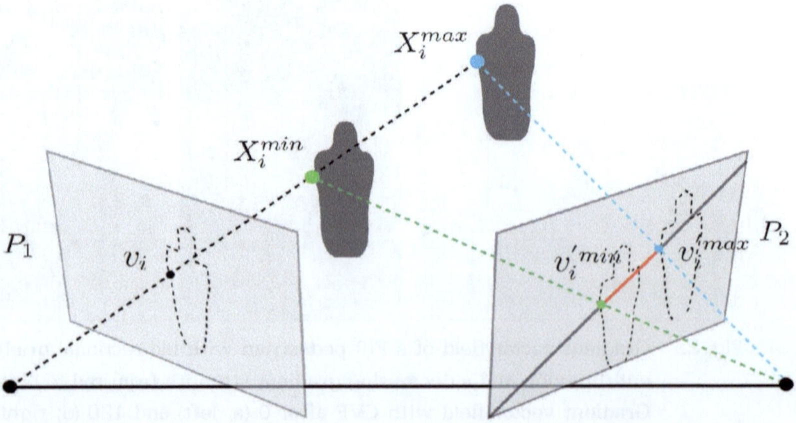

Fig. 3. Left: Projection P_1 from X_i^{min}, X_i^{max} to v_i. Right: Projection P_2 from X_i^{min}, X_i^{max} to $\mathbf{v}_i^{'min}$ and $\mathbf{v}_i^{'max}$. The new restricted search space at the epipolar line $l_i^{'}$ is color-marked.

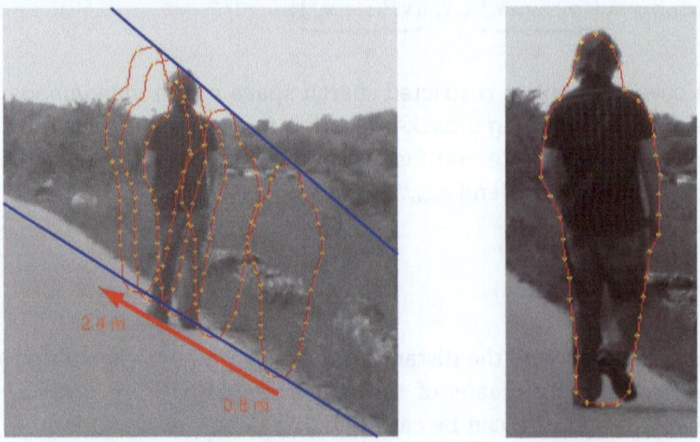

Fig. 4. (a, left) Multiple projections of the active contour model into the video image. The height of the model has been scaled within the interval $[0.8 \ldots 2.4]$ m. The distance has been increased with a step width of 0.4 m. The epipolar lines for the head and the foot point are marked in blue colour. (b, right) Model $\mathscr{M}_{opt}$ with the smallest error.

In a last step the best fitting projected active contour model $\mathbf{V}_{opt}$ in respect to the edge information has to be selected out of $\mathscr{M}_{search}$. If the correct corresponding model in the video image has been found, the distance of the pedestrian can be deduced via triangulation. As a result of noise and imprecise calibration the resulting lines are skewed and do not fullfil the epipolar constraint. For that reason the point with the smallest distance to the two triangulation lines has to be found. This can be achieved by solving a minimization problem with a singular value decomposition [13]. The final distance result is the average of all triangulated active contour model points.

3 Classification

The image region which is used as initial starting area for the active contour stereo approach is selected by a hypotheses generation process which focuses on hot spots in the far infrared image. The selection is performed in a rule-based manner which solely utilizes simple edge relationships as hypothesis indication. Thus, both the false positive and the true positive rate are rather high. In order to discard as many non-pedestrian hypotheses as possible a subsequent classification step is essential.

The feature set of the current state-of-the-art pedestrian detector from Dalal and Triggs [14] is based on normalized Histograms of Oriented Gradient (HOG) descriptors. They are similar to the SIFT image descriptors [15] but are computed on a dense grid of uniformly spaced cells. An image with pixel size $w \times h$ which is split up into histograms H_i with n gradient bins and a rectangular cell size of k pixels results in the feature vector

$$\mathbf{v} = (H_1 = (h_1, \ldots, h_n), H_2, \ldots, H_{\frac{w \cdot h}{k^2}}) \tag{3}$$

A L2-normalization of the histograms is necessary to compensate for additive and multiplicative illumination variations. Dalal achieved the best results with a normalization which also accounts the cell neighbours. In order to preserve real-time performance we decided to use a normalization without considering the cell-context:

$$\tilde{H}_i = \frac{H_i}{\sqrt{\sum_{k=1}^{n} h_k^2}} \tag{4}$$

Finally, a support vector machine with linear kernel is trained with the feature vectors $\tilde{\mathbf{v}}_i$ which are extracted from a sufficient amount of pedestrian and non-pedestrian image patches.

The algorithm performs best if the pedestrian is image-centered with an additional small pixel margin around. To ensure that, the best-match model $\mathcal{M}_{opt}$ is chosen to calculate its rectangular bounding box. An extra margin relative to the width and height of the bounding box is added to the region. The image region is scaled with a linear filter to match the pedestrian model size of 128x64. This has to be done, since the feature extraction operates solely on equally-sized image patches.

4 Sensor Configuration

We have set up an experimental car with a CMOS video camera and a far infrared imaging device. The video camera is situated behind the wind shield on the front passenger side while the far infrared imaging device has been integrated into the bumper on the drivers side. These sensor mounting points have been chosen to ensure an almost maximal possible camera-baseline for an automotive front-percepting stereo-system (see Fig. 5). This leads to an increased spatial accuracy in higher distances.

Fig. 5. Mounting locations of the video camera (top) and the far infrared image device (bottom) on the experimental car.

5 Results

A first evaluation was performed on several outdoor scenarios under different weather conditions (see Fig. 6, 7 and 8). All FIR pedestrian positions were manually labeled [16] and serve as initial starting point for the active contour model energy minimizing process. The scenes were recorded with constant own car velocities and the pedestrians walked likewise in a constant manner. Therefore, the measured distances between experimental car and located pedestrian should also vary linearly.

The measurement results of the 3D reconstruction are visualized in the scatter plot diagrams. False active contour matches are highlighted with a circle. The red regression line shows the linear behavior of the distance values. This procedure does not need ground truth data of the pedestrian positions. However, a precise accuracy estimation is not possible with this course of action. Exact ranging measures with a reference sensor such as a laser scanner would be absolutely essential to determine measuring variances. The used evaluation method offers only a rough performance estimation.

We achieve real-time capabilities on a dual core processor with 2.16 GHz as long as only one model has to be calculated. The calculation times split up as follows: Approximately 10% are needed for the calculation of the gradient images, 55% for the calculation of the active contour model including the retrieval of the corresponding model in the video image, 15% for the triangulation and reconstruction and 20% for the classification. The calculation time sums up to 24 ms per model on average.

Fig. 6. Scene 1 was recorded at rainy weather on a testing site. The distance of the pedestrian which has been measured with the ACSA algorithm is 26.8 m in the FIR and video images (a, upper left), (b, upper right). (c, bottom) shows the measured distances in comparison to the linear regression (red line) while approaching the pedestrian.

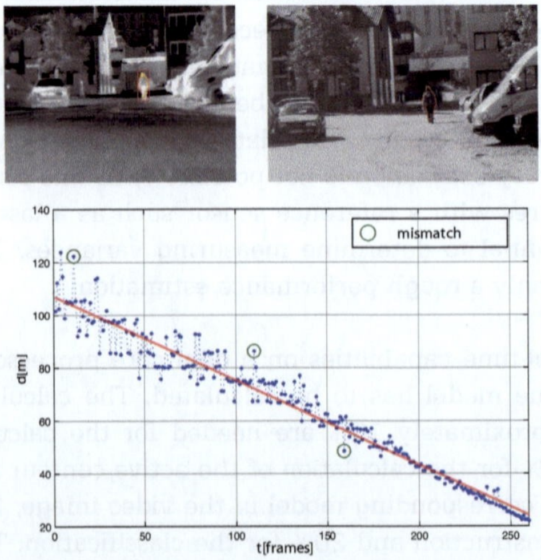

Fig. 7. was recorded in the urban area. Warm background objects are visible. The distance of the pedestrian in the FIR (a, upper left), and the video (b, upper right), image is 33.7 m. (c, bottom) shows the measured distances in comparison to the linear regression (red line) while approaching the pedestrian

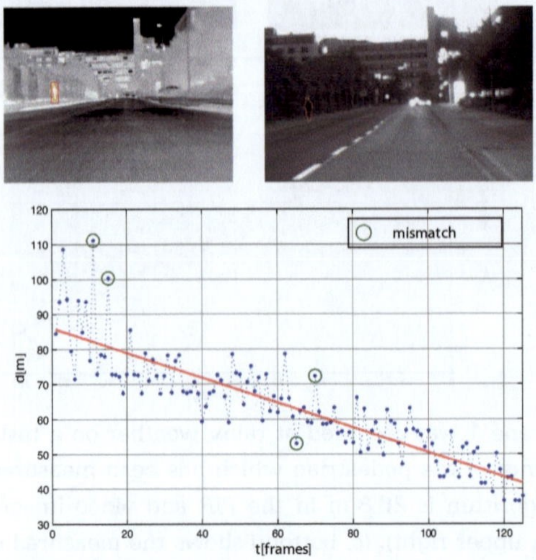

Fig. 8. was recorded in the urban area epsdawn. The distance of the pedestrian in the FIR (a, upper left), and the video (b, upper right), image is 43.5 m. (c, bottom) shows the measured distances in comparison to the linear regression (red line) while approaching the pedestrian.

6 Conclusions and Future Work

6.1 Conclusions

The paper at hand proposes a fusion method of multi spectral video sources for pedestrian localization. Since the nature of a FIR image is quite different from a standard video image one of the biggest challenges is to find useful information that can be combined from both imagers. A new approach for stereo-matching, based on contour information as common feature, is introduced. In the first step the object contour is extracted by using extended active contour models with gradient vector flow. In the second step the stereo correspondence problem is solved with a fast active contour shape matching algorithm utilizing the epipolar constraint. Finally, a postponed multi-spectral classification confirms or discards the object.

6.2 Future Work

There are some possibilities for additional methods to improve the convergence of the active contours. Whenever the approximation of the pedestrian contour does not fit well, the correct determination of the model position in the video image applying the presented cost-function is not determinable. The resulting effect is, that consecutively viewed position results can vary a lot in the distance assigned to the reconstructed 3D-position. Improvements in the accuracy of the pedestrian contour alignment should result in a higher precision shape matching in the video image. The last but most important point is the evaluation of image sequences in the corresponding FIR- and video images.

Currently, the multi spectra stereo reconstruction is performed frame by frame. Tracking methods incorporating driving dynamic, would improve the object stability, even when consecutive classification results end up in false negative matches. Several suitable tracking algorithms are conceivable [17, 18]. An object classified at time step t could be reassigned at time step $t+1$ with a certain probability. Computing time can be reduced, whenever it is possible to use a preformed initial active contour model from the preceding step. Hence, the number of convergence iterations could be reduced significantly, because there is no need to use the generalized ellipse as initial model. Moreover, computational load can be significantly reduced by applying the distance information derived from the 3D-reconstruction process. If a pedestrian can be matched to its preceding match the distance information from the 3D reconstruction can be added to the matching process and the search space could be greatly reduced.

Finally, further evaluation has to be performed to acquire both recognition rates and accuracy on a more versatile training set. A reference sensor system to acquire ground truth information is mandatory.

References

[1] M. K. A. W. D. Terzopoulos, "Snakes: Active contour models," CCV, pp. 259–268, 1987.

[2] D. Terzopoulos, J. Platt, A. Barr, and K. Fleischer, "Elastically deformable models," SIGGRAPH Comput. Graph., vol. 21, no. 4, pp. 205–214, 1987.

[3] L. C. S. W. W. C., "Supervised multispectral image segmentation using active contours," Robotics and Automation, 2005. ICRA 2005. Proceedings of the 2005 IEEE International Conference, pp. 4242– 4247, 2005.

[4] D. Terzopoulos and R. Szeliski, "Tracking with kalman snakes," pp. 3–20, 1993.

[5] Z. Hou and C. Han, "Force field analysis snake: an improved parametric active contour model," Pattern Recognition Letters, vol. 26, no. 5, pp. 513–526, 2005.

[6] A. P. J. Boyce, "An implementation of the active contour method for noisy images using a local minimisation algorithm," Robotics and Digital Technology Monash University Clayton VIC 3168 Australia, Tech. Rep., 1995.

[7] I. Cohen, L.D.; Cohen, "Finite-element methods for active contour models and balloons for 2-d and 3-d images," Pattern Analysis and Machine Intelligence, vol. Volume 15, pp. 1131–1147, Nov 1993.

[8] J. L. P. Chenyang Xu, "Gradient vector flow: A new external force for snakes," IEEE Computer Society Conference on Computer Vision and Pattern Recognition (CVPR'97), p. 66, 1997.

[9] C. Xu and J. L. Prince, "Snakes, shapes, and gradient vector flow," IEEE Transactions on Image Processing, vol. 7, pp. 359–369, Mar. 1998.

[10] C. Scott and R. Nowak, "Robust contour matching via the order-preserving assignment problem," Image Processing, IEEE Transactions on, vol. 15, no. 7, pp. 1831–1838, 2006.

[11] S. V. Graham McNeill, "Part-based probabilistic point matching," 18th International Conference on Pattern Recognition (ICPR'06), pp. 382–386, 2006.

[12] L. X. J. Y. W. Yanjie, "An eigenvector approach based on shape context patterns for point matching," Communications and Information Technologies, 2006. ISCIT '06. International Symposium on, pp. 455–458, 2006.

[13] R. I. Hartley and A. Zisserman, Multiple View Geometry in Computer Vision, 2nd ed. Cambridge University Press, ISBN: 0521540518, 2004.

[14] B. Dalal, N. Triggs, "Histograms of oriented gradients for human detection," Computer Vision and Pattern Recognition, 2005. CVPR 2005. IEEE Computer Society Conference on, vol. 1, pp. 886– 893, Jun. 2005.

[15] D. G. Lowe, "Object recognition from local scale-invariant features," iccv, vol. 02, p. 1150, 1999.

[16] T. Tatschke, F. Färber, E. Fuchs, L. Walchshäusl, and R. Lindl, "Semi-autonomous reference data generation for perception performance evaluation," in Proceedings of the 10th International Conference on Information Fusion (FUSION), Quebec, July 2007.

[17] A. M. Baumberg and D. C. Hogg, "An efficient method for contour tracking using active shape models, Tech. Rep. 94.11, April 1994. [Online]. Available: citeseer.ist.psu.edu/baumberg94efficient.html.

[18] G. S. M. Bertalmio and G. Randall, "Morphing active contours," IEEE PAMI, vol. 22, July 2000.

Günter Bauer, Florian Homm, Leonhard Walchshäusl
BMW Group Research and Technology
Hanauer Straße 46
80992 Munich
Germany
guenter.bauer@bmw.de
florian.homm@bmw.de
leonhard.walchshaeusl@bmw.de

Darius Burschka
Technische Universität München
Fakultät für Informatik
Boltzmannstr. 3
85748 Garching
Germany
burschka@cs.tum.edu

Keywords: pedestrian localisation, multi spectral stereo, active contour model

Car Driver Monitoring by Networking Vital Data

J. Murgoitio, Robotiker: Tecnalia-Automoción
J. I. Fernández, ESIDE: University of Deusto

Abstract

According to the "Ambient Intelligence" concept (AmI), Tecnalia-Automoción is working on several new applications and one of them is an ECG sensor integrated with the steering wheel. A previously developed prototype of a sensorized active headrest which was designed to maintain desired horizontal and vertical safety distances to head is related to this work.

Even though no continuous ECG signal could be acquired from the steering wheel because hands position is changing, nevertheless it would be possible to sample this information and find out how some particular parameters are changing during the driving activity, like those based on HRV (Heart Rate Variability). Then, Tecnalia-Automoción is designing a networked solution with smart sensors and actuators integrating the IEEE 1451 standard group, to be applied to the previously mentioned active headrest prototype and this ECG sensor.

1 Introduction

AmI is the abbreviation for the "Ambient Intelligence" concept, the selected words by ISTAG (Information Society Technologies Advisory Group) as a guiding vision to give an overall direction to Europe's Information Societies Technology programme. AmI stresses the importance of social and human factors as well as developing the base technologies on which aspects of the vision are founded, and it's the limit of a process which introduces the technology into people's lives in such a way that the introduction never feels like a conscious learning curve: no special interface is needed because human experience is already a rich 'Manual' of ways of interfacing to changing systems and services.

According to these "AmI" principles, Tecnalia-Automoción is working on several new applications. One of them is a developed prototype of a sensorised active headrest which it is shown within, the "Previous works" chapter. This application defins some common solutions to network several sensors and actuators for other different AmI applications.

The number of car sensors and actuators is increasing (an average car has between 50 and 100 sensors). This means that solutions related to networking sensors and more intelligent sensors (more features on the sensor side) are being considered as strategic ways for the future, not only for sensors but also for actuators as elements to execute actions within a distributed control system. Tecnalia-Automoción is researching intelligent sensor networks and his availability and application in car systems and devices. All sensors and actuators mentioned would be integrated within a networked system taking the IEEE 1451 standard for a "plug & play" system into account.

In the following paragraphs is described how these concepts and standards considered for the active headrest are being applied to an ECG sensor.

2 Previous Works

One of the previously mentioned applications is a developed prototype of a sensorized active headrest which was designed to maintain desired horizontal and vertical safety distances to driver head.

This active headrest detects the position of the user's head using measured information by contact-less devices: two infrared sensors placed in the headrest. One of them is a presence detecting sensor and the other one is an analogical sensor to measure the distance to an object. With these, we can pinpoint the edge of the head and find out the distance from the headrest to the head. The system adjusts the headrest to the optimal safety position using two independent motion controls for the horizontal and vertical directions. A clarifying scheme is shown in Fig. 1.

All sensors (4) and actuators (2) of this application were designed to be integrated within a networked system and taking the IEEE 1451 standard for a "plug & play" system into account.

The following paragraphs describe how concepts and standards related to intelligent sensor networks are applied to an ECG sensor too.

3 ECG Sensor: State of the Art

Two different anatomical structures are used as physiological indicators of workload measures: Central Nervous System (CNS, it includes the brain, brain stem, and spinal cord cells), and Peripheral Nervous System (PNS). The PNS can be divided into the Somatic Nervous System (SoNS), which is concerned with the activation of voluntary muscles and Autonomic Nervous System (AuNS), which controls internal organs and is autonomous because AuNS innervated muscles are not under voluntary control. The AuNS is further subdivided into the Parasympathetic Nervous System (PaNS) maintaining bodily functions and the Sympathetic Nervous System (SyNS) for emergency reactions.

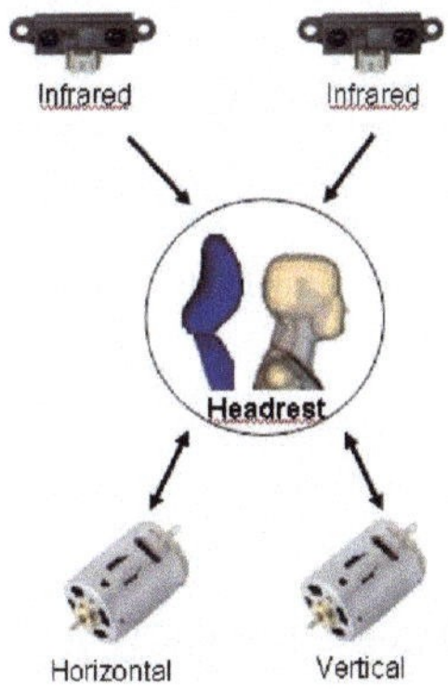

Fig. 1. Active headrest

Most organs are dually innervated both by SyNS and PaNS, and both can be coactive, reciprocally active, or independently active. Heart rate is controlled by the AuNS.

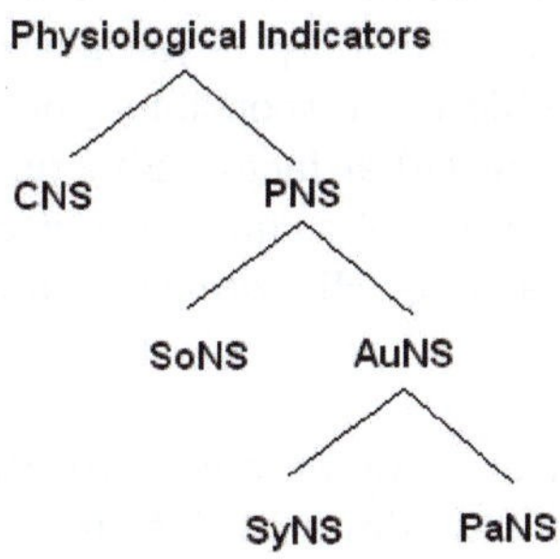

Fig. 2. Anatomical structures

3.1 Heart Activity

As mentioned before, the heart is innervated both by the PaNS and SyNS. Each heart contraction is produced by electrical impulses that can be measured in the form of the ECG (Electrocardiogram). The following figure shows typical register of heart electrical activity:

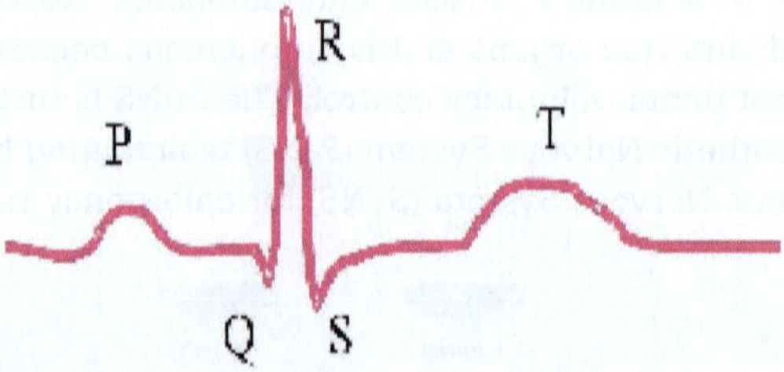

Fig. 3. Heart activity

Time domain, frequency and amplitude measures can be derived from the ECG signal:

Time Domain:

In the time domain the R-Waves of the ECG are detected, and the time between these peaks (IBI: Inter Beat Interval) is calculated. IBI is directly related to Heart Rate (HR), however, this relation is non linear and IBI is more normally distributed in samples compared with HR. Therefore, IBI scores should be used for detection and testing of differences between mean HR. IBI scales is less influenced by trends than the HR scale.

According to some scientific works, average heart rate during task performance compared to rest-baseline measurement is a fairly accurate measure of metabolic activity, and not only physical effort affects heart rate level. Emotional factors, such high as responsibility or the fear of failing a test, also influence mean heart rate. Other factors affecting cardiac activity are speech and high *G*-forces. The effect of sedative drugs and time-on-task resulting in fatigue is a decrease in average HR, while low amounts of alcohol are reported to increase HR.

In the time domain, heart rate variability (HRV) is also used as measure of mental load. HRV provides additional information to average HR about the feedback between the cardiovascular systems and CNS structures. In general HRV decrease is more sensitive to increases in workload than HR increase, although there have been several reports of both HR and HRV unaffected. One of the causes for finding no effect of mental load on HRV lies in the globalness of the measure and its sensitivity to physical load. Some works showed that an increase in physical load decreased HRV and increased HR, while an increase in mental load was accompanied by a reduced HRV and no effect on HR (Lee & Park, 1990). Fatigue is reported to increase HRV (Mascord & Heath, 1992) while low amounts of alcohol decrease HRV (González González et al., 1992).

Frequency Domain:

In frequency domain, HRV is decomposed into components that are associated with biological control mechanisms (Kramer, 1991; Porges & Byrne, 1992). Three frequency bands have been identified (L.J.M. Mulder, 1988, 1992): a low frequency band (0.02-0.06 Hz) believed to be related to the regulation of the body temperature, a mid frequency band (0.07-0.14 Hz) related to the short term blood-pressure regulation and a high frequency band (0.15-0.50 Hz) believed to be influenced by respiratory-related fluctuations (vagal, PaNS influenced – Kramer, 1991):

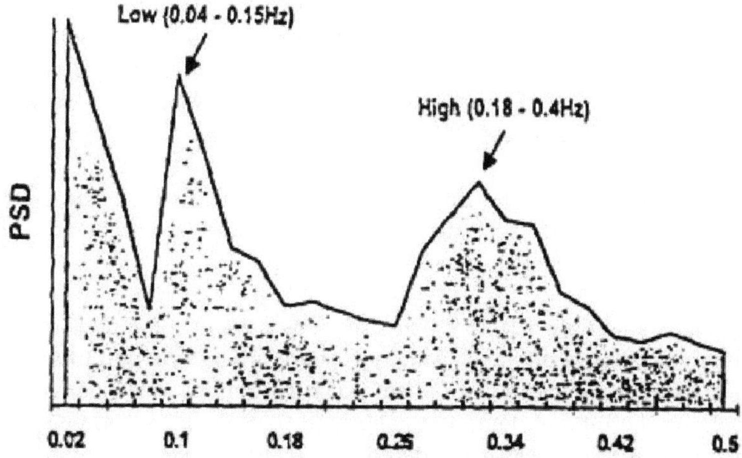

Fig. 4. HRV: frequency analysis (PSD = Power Spectral Density). (The John D. & Catherine T. MacArthur Foundation)

A decrease in power in the mid frequency band ("0,10 Hz" component) and in the high frequency band have been shown to be related to mental effort and task demands (Jorna, 1992; Backs & Seljos, 1994; Paas et al.-1994).

Amplitude Domain:

Finally, amplitude information from the ECG signal can be used to obtain information about workload. The amplitude of the T-wave (TWA) is said to mainly reflect SyNS (Furedy, 1996) and decreases with increases in effort.

3.2 Driver mental workload

The reason to consider the integration of driver ECG signal within the car system is closed related to the "driver mental workload" measurement and the relation with some physiological indices. One of them is the heart activity and the changes produced on some parameters calculated from the ECG signal.

Driving is a very dynamic task in a changing environment. Moreover, the driving task is large influenced by drivers themselves. Nowadays, there are factors that may even lead to increased human failure in traffic:

- The number of vehicles on the road is increasing, so increased road intensity leads to higher demands on the human information processing system and an increased likelihood of vehicle collisious.
- People continue to drive well into old age. Elderly people suffer from specific problems in terms of divided attention performance, a task that is more and more required in traffic. One of the causes of these increased demands is the introduction of new technology into the vehicle.
- Drivers in a diminished state endanger safety on the road (longer journeys, night time driving, and so on). Driver fatigue is currently an important factor in accident causation.

The above mentioned factors and situations have in common that in all cases driver workload is affected. Although there are several definitions and models to explain it, "mental workload" could be defined as a relative concept; it would be the ratio of demand to allocated resources. From this point of view, several scientific works have demonstrated that some parameters obtained from physiological measures (pupil diameter, heart rate and respiratory, electro dermal activity, EEG, electro-oculography, ...) could help to know the driver´s mental workload and one of them is the ECG. Due to its low level invasive characteristic, ECG information seems very interesting information to increase safety in driving tasks. Main idea is to use laboratory methods considered in traffic research and based on ECG signal, and integrate these in a car to improve safety and/or wellness.

4 Adopted Solution

Measurement of heart rate is not very complex, the ECG signal needs little amplification (about 10 to 20 times less as EEG) and electrode placement is not very critical if measurement is limited to R-wave detection and registration. Then, the following solution has been selected by Tecnalia-Automoción to integrate an ECG sensor in a car:

So no continuous ECG signal could be acquired from the steering wheel because hands position is changing, nevertheless it would be possible to sample this information and find out how some particular parameters are changing during the driving activity. According to this schema, heart rate (HR) may provide an index of overall workload, and spectral analysis of heart rate variability (HRV) would be more useful as index of cognitive, mental workload (Wilson & Eggemeier, 1991).

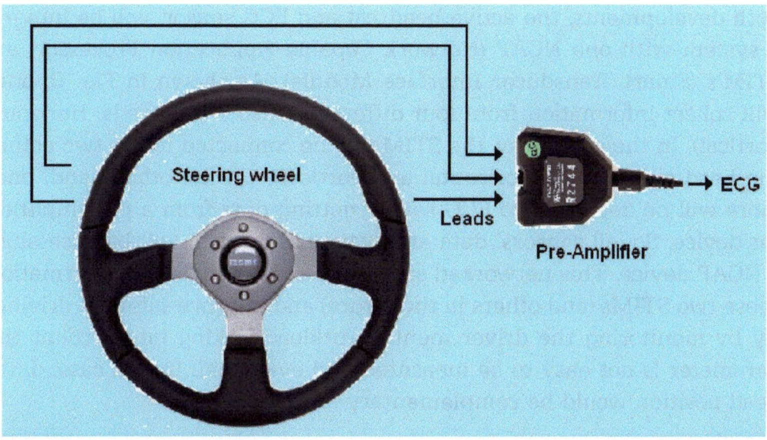

Fig. 5. Adopted solution.

4.1 Networking

Tecnalia-Automoción is designing a networked solution with smart sensors and actuators, to be applied to the previously mentioned active headrest prototype and this ECG sensor.

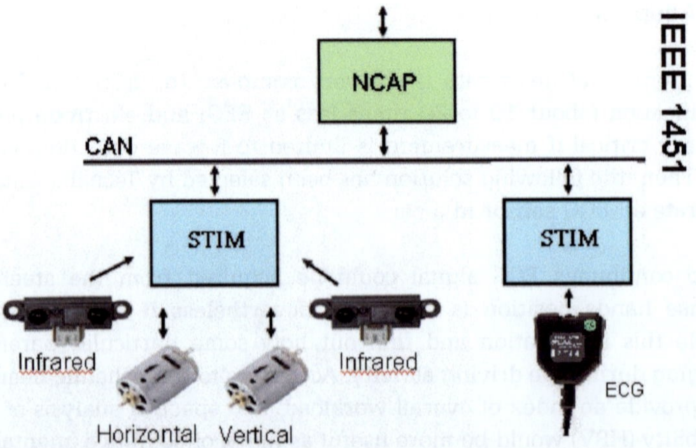

Fig. 6. Networked solution.

Both developments, the active headrest and ECG sensor, will be integrated in a system with one NCAP (Network Capable Application Processor) and two STIM's (Smart Transducer Interface Module). As shown in Fig. 6, one STIM will collect information from four different sensors (Infrareds, Horizontal and Vertical). In the same way, the STIM will be connected to the two actuators of the head-rest system (horizontal and vertical). On the other hand, one STIM more will be dedicated to ECG sensor getting data from a pre-amplifier & filter device. So, all sensors' data and actuators' control will be accessible from a NCAP device. This networked strategy will let to combine information from these two STIMs (and others in the future) and improve safety in driving activity by monitoring the driver mental workload, taking into account that this parameter is not easy to be measured and evaluated. In this case, data about head position would be complementary and welcome.

4.2 Plug & Play Feature

Sensors and actuators for both AmI developments (active headrest and ECG) will be integrated within a networked system taking into account the IEEE 1451 standard to add "plug & play" feature.

The family of IEEE 1451 is a group of seven standards, some of them in revision phase, describing a set of open, common, network-independent communication interfaces for connecting transducers (sensors or actuators) to microprocessors, instrumentation systems, and control networks; and to make it easier for transducer manufacturers to develop smart devices and to interface these devices to networks, systems, and instruments. The goal of 1451 is to allow the access of transducer data through a common set of interfaces whether the transducers are connected to systems via a wired or wireless means.

Only the following two standards from the IEEE 1451 group have been considered in this work:

IEEE P1451.0: providing a uniform set of commands and TEDS (Transducer Electronic Data Sheet) for the family of IEEE 1451 smart transducer standards. This command set let to access any sensors or actuators in the 1451-based networks. This standard will be used to assure uniformity within the family of IEEE 1451.x interface standards.

IEEE P1451.6: defining a transducer-to-NCAP interface and TEDS using the high-speed CANopen network interface. It defines a mapping of the 1451 TEDS to the CANopen dictionary entries as well as communication messages, process data, configuration parameter, and diagnosis information.

TEDS definition is the key feature of this family of standards and it would be a memory device attached to the transducer having information like transducer identification, calibration, correction data, measurement range, manufacture-related information, and so on.

A general overview for the ECG sensor STIM is shown in Fig. 7:

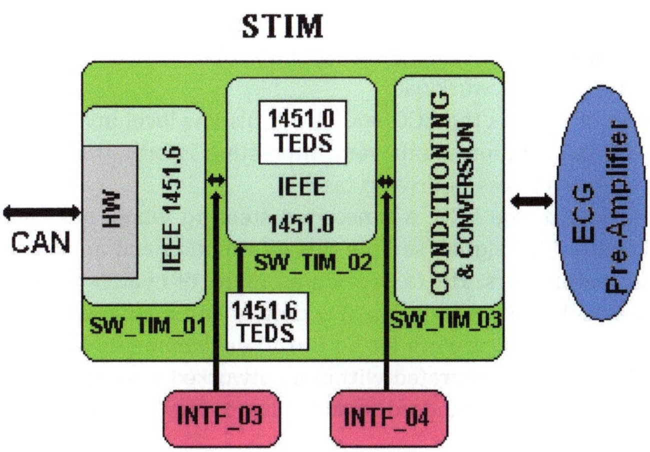

Fig. 7. STIM.

STIM has been designed considering tree modules working together too. SW_TIM_01 will be an implementation of IEEE 1451.6 standard (STIM side). SW_TIM_02 will have implemented the 1451.0 standard and will manage

TEDS with all information related to the ECG sensor connected to this STIM. Finally, SW_TIM_03 will do conditioning tasks for data received from ECG Pre-Amplifier. As it's shown in the previous figure, two more interfaces are considered (INTF_03 and INTF_04). This STIM device will be similar device for the active headrest but only modifying his TEDS data according to his connected sensors and actuators.

5 Conclusions

Summarizing all activities carried out in this work, the following conclusions could be underlined:

- The number of car sensors and actuators is increasing, it means that solutions related to networking sensors and more intelligent sensors are being considered as strategic ways for the future.
- Tecnalia-Automoción is researching intelligent sensor networks for AmI systems and his application in cars: A networked solution formed by an active headrest and an ECG sensor is being developed.
- The driving task is very influenced by drivers themselves and there are factors that may even lead to increased human failure in traffic. These factors and situations have in common that in all cases driver mental workload is affected.
- Some scientific works have demonstrated that some parameters obtained from physiological measures (pupil diameter, heart rate and respiratory, electro dermal activity, EEG, ...) could help to know the driver mental workload.
- One of them is the ECG and due to his low level invasive characteristic, ECG information seems very interesting information to increase safety and/or wellness in driving tasks.
- Related to the ECG, a sensorized steering wheel has been adopted to acquire ECG signal, sample this information and analyse some particular parameters: HR to provide an index of overall workload, and HRV spectral analysis as index of cognitive and mental workload.
- Sensors and actuators for both AmI developments (active headrest and ECG) will be integrated within a networked system taking into account the IEEE 1451 standard to add "plug & play" feature.

References

[1] Backs, R.W.; Seljos, K.A. (1994). Metabolic and cardio respiratory measures of mental effort: the effects of level of difficulty in a working memory task. International journal of Psychophysiology, 16, 57-68.

[2] De Waard, Dick (1996). The measurement of driver's mental workload. [Phd Thesis]. University of Groningen, June 1996.

[3] Furedy, J.; Szabo A.; Péronnet, F. (1996). Effects of psychological and physiological challenges on heart rate, T-wave amplitude, and pulse-transit time. International journal of psychophysiology, vol 22, 173-183.

[4] González González, J.; Mendez Llorens, A.; Mendez Novoa, A.; Cordero Valeriano, J.J. (1992) Effect of acute alcohol ingestion short term heart rate fluctuations. Journal of studies on Alcohol, 53, 86-90.

[5] Gorman, Bryan (2005). Towards a Standards-Based Framework for Interoperable CBRN Sensor Networks . [Conference]. Hampton-Virginia: SensorsGov Expo & Conference, USA, December 2005.

[6] Jorna, P.G.A.M. (1992). Spectral analysis of heart rate and psychological state: a review of its validity as a workload index. Biological Psychology, 34, 237-257.

[7] Kramer, A.F. (1991). Physiological metrics of mental workload: a review of recent progress. In D.L. Damos (Ed.), Multiple-task performance. (pp 279-328).

[8] Lee, Kang (2005). Synopsis of IEEE 1451. [Conference]. Chicago: Sensors Conference/Expo 2005, USA, June 2005.

[9] Lee, D.H.; Park, K.S. (1990). Multivariate analysis of mental and physical load components in synus arrhythmia scores. Ergonomics, 33, 35-47.

[10] Mascord D.J.; Heath (1992). Behavioural and physiological indices of fatigue in a visual tracking task. Journal of safety research, 23, 19-25.

[11] Mulder, L.J.M. (1988). Assessment of cardiovascular reactivity by means of spectral analysis. PhD Thesis. Groningen: University of Groningen.

[12] Mulder, L.J.M. (1992). Measurement and analysis methods of heart rate and respiration for use in applied environments. Biological Psychology, 34, 205-236.

[13] Paas, F.G.W.C.; Van Merrienboer, J.G.J; & Adam, J.J. (1994). Measurement of cognitive load in instructional research. Perceptual and motor skills, 79, 419-430.

[14] Porges, S.W.; Byrnes, E.A. (1992). Research methods for measurement of heart rate and respiration. Biological Psychology, 34, 93-130.

[15] Wiczer, James; Lee, Kang (2005). A Unifying Standard for Interfacing Transducers to Networks – IEEE-1451.0 . [Workshop]. Lviv: IEEE International Workshop on Intelligent Data Acquisition and Advanced Computing Systems: Technology & Applications, Ukraine, 8-10 September.

[16] Wilson, G.F.; Eggemeier, F.T. (1991). Psycho physiological assessment of workload in multi-task environments. In D.L. Damos (Ed.), Multiple-task performance (pp. 329-360). London: Taylor & Francis.

J. Murgoitio
Tecnalia-Automoción
Parque Tecnológicomódulo 202
48170 ZAMUDIO - Bizkaia
Spain
murgoiti@robotiker.es

J. I. Fernández
ESIDE - University of Deusto
Avenida de las universidades 24
48007 BILBAO - Bizkaia
Spain
jguti@eside.deusto.es

Keywords: AmI, automotive, mental workload, ECG, sensor network, plug&play, IEEE1451

MDSI Range Camera Calibration

T. Hanning, University of Passau
A. Lasaruk, FORWISS University of Passau
R. Wertheimer, BMW Group Forschung und Technik

Abstract

An MDSI (Multiple Double Short-Time Integration) range camera consists of a grid of pixels which are utilized to measure the near infrared laser intensity back-scattered from an illuminated scene. For each pixel two consecutive intensity measurements are conducted which encode the time-of-flight as well as the reflectance of the illuminated objects. From these two intensity values the distance to the object in the observed solid angle element can be computed. Estimation of range camera parameters which are essential for accurate range reconstruction can be performed using a coordinate-measuring device. The corresponding calibration procedure, however, is tedious and inflexible. In this paper we therefore present a new, simple and flexible approach to range camera calibration based on separate representations of viewing rays corresponding to the range pixels. Our approach uses range reconstruction superiority of a calibrated gray-value camera for planar calibration patterns. The resulting range information is given in the coordinate system of the gray-value camera.

1 Introduction

To keep safe distances to obstacles and road users such as motor vehicles, cyclists, and pedestrians, one needs reliable depth information to relevant objects in the front scenario. Such range information can be obtained at low cost using homogeneous multi-sensor systems combined of a matrix of time-of-flight range sensors. Prominent examples of such sensors are laser scanners and range cameras like the 64x8 range camera developed within IP PReVENT, subproject UseRCams, led by Siemens CT.

A range camera consists of a grid of range pixels. Each range pixel provides the distance between the observed scene and the sensor element. In a first approximation, this distance can be idealized as being measured along a viewing ray. A feasible way to realize a range camera, which is even suitable for outdoor measurements, has been presented by Mengel et. al. [1, 2]. In their MDSI (Multiple Double Short-Time Integration) approach each range pixel integrates the radiation emitted by a near infrared laser and reflected by the observed objects during two time periods of different length. For each range measurement two intensity acquisitions are made, starting consecutively at time t_1 and t_2, respectively. For each intensity measurement a separate laser pulse is emitted. Let the radiation intensity at area element $a \in A$, where A is the active sensor area, and time t be denoted by $\Phi(a, t)$ and let the sensor sensitivity at point a be denoted by $\Psi(a)$. Then the (voltage) signals s and l corresponding to the two integration measurements are given by the following expressions:

$$s = \int_{t_1}^{t_1+t_s} \int_A \Phi(a,t)\, \Psi(a)\, da\, dt \qquad l = \int_{t_2}^{t_2+t_l} \int_A \Phi(a,t)\, \Psi(a)\, da\, dt \tag{1}$$

(These equations hold, provided that the movement of the scene objects is negligible - which implies that the reflectivity assigned to the observed solid angle element remains constant during the two subsequent integration periods.)

Both acquisition periods start before the backscattered light from the associated laser pulse reaches the sensor surface. It is assumed that the inequality $t_1+t_s<t_2$ holds, although the difference is very small. The short shutter period t_s is chosen in such a way that s contains a part of the laser radiation reflected from the scene. This fraction depends on the round-trip run-time of the laser pulse. The second shutter period t_l is significantly longer than t_s; hence within the period l all laser pulse radiation reflected from the solid angle element of the scene is collected.

According to [1] the distance to the observed object is linear in the quotient s/l. In other words, for each range pixel there exist coefficients p,q, such that for the distance to the observed scene $d(s,l)$ the following holds:

$$d(s,l) = p\left(\frac{s}{l}\right) + q \tag{2}$$

A range camera calibration is conceived as an algorithm, which takes as input for each range pixel a set of signal pairs $(s_1, l_2),\ldots,(s_n, l_n)$ obtained from observed objects of known nature. It returns the parameters p and q for each range pixel.

Finding a simple way how to obtain the parameters p and q turns out to be a challenging task. The measured signals are subject to strong noise. Furthermore, both signal values are perturbed by ambient influences, like reflectivity of the observed scene, saturation of the hardware units, or crosstalk effects. A prevalent tool to perform range calibration is a coordinate measuring device. However, the corresponding hardware setup is inflexible, expensive, and time-consuming. Furthermore, such a calibration setup suffers from the inherent error implied by the unknown relation between the coordinate system of the range camera and that of the coordinate-measuring device.

A task equivalent to range calibration as defined above is to establish a function, which for two corresponding intensities provides the Z coordinate $z(s,l)$ of the observed world point in some user-defined reference coordinate system rather than returning just the distance $d(s,l)$ to the sensor surface pixel. In automotive applications a suitable reference coordinate system may be defined with regard to a front-view gray-value camera. Since the estimation of the spatial relation between the range and gray-value camera is difficult, it is sensible to represent the range information directly in the coordinate system of the gray-value camera. Many applications of range cameras, like segmentation of objects, can even be better performed using the Z coordinate instead of the distance measure. For this reason, we speak of range calibration also in the case of estimation of the function z.

In this paper we present a new simple and fully autonomous range calibration procedure for an MDSI range camera which is simple to execute and can be conducted largely autonomously. Our method is based on range reconstruction superiority of a calibrated gray-value camera for known observed objects. The coordinate-measuring device is replaced in our approach by a gray-value camera and a plate with a suitable pattern. Since the accuracy of calibration plate reconstruction utilizing the gray-value camera is considerably higher than of the range camera, the proposed procedure provides precise and stable range camera calibration with respect to the coordinate system of the gray-value camera. It is one result of this paper that pixel-wise range camera distance calibration and Z coordinate calibration are exchangeable: The obtained parameters can be transformed into each other with little additional calibration effort. Our contribution is a first step on the way to a full calibration of an MDSI range camera: A fully calibrated MDSI range camera will not only provide distance data but also coordinate data.

The outline of the paper is as follows: In Section 2 the calibration setup and the calibration procedure from the perspective of the user are described. Section 3 deals with the details of the calibration procedure and the corresponding mathematical background. In Section 4 we give a sketch of our experimental results and we discuss the accuracy of the proposed method. Section 5 summarizes the achieved results and gives an outlook about future development efforts.

2 Hardware Setup

The hardware setup consists of an off-the-shelf gray-value camera and an MDSI range camera with rigid mutual spatial relation. It has to be emphasized that once the system has been calibrated the special relation between the gray-value camera and the range camera must remain fixed. An obvious requirement is to keep the effects of the displacement parallax small. This implies that the relevant fields of views (FoV) of both cameras overlap and that their positions do not differ too much when compared to the distances of interest.

A sample calibration setup is depicted in Fig. 1. The gray-value camera must be calibrated first, for example by using the methods described in [3].

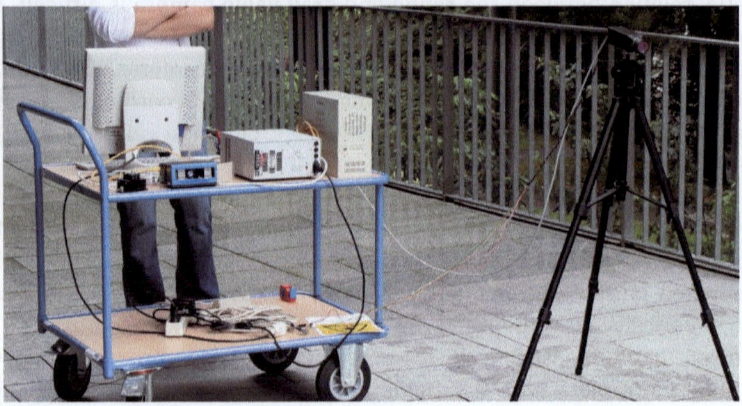

Fig. 1. Hardware setup

The planar reference calibration pattern can be chosen mostly arbitrarily: It is even possible to use different calibration patterns for different distance ranges. The only requirement is that the calibration pattern can be used to reconstruct the plane equation of the pattern surface in the gray-value camera coordinate system. Several methods how to achieve this are described in [4]. One possible pattern is depicted in Fig. 2.

During image sequence acquisition, the pattern is continuously exposed at different distances and orientations to both cameras. From these sequences, only images of plane orientations orthogonal to the optical axis of the gray-value camera are selected: Appropriate choices can be made by the requirement that the parameters of each reconstructed plane Eq.

$$ax + by + cz = d \tag{3}$$

satisfy the following constraints:

$$\langle (a,b,c) \| (1,0,0) \rangle = 0, \quad \langle (a,b,c) \| (0,1,0) \rangle = 0. \tag{4}$$

This is achieved by using stable geometric reasoning methods as described in [5].

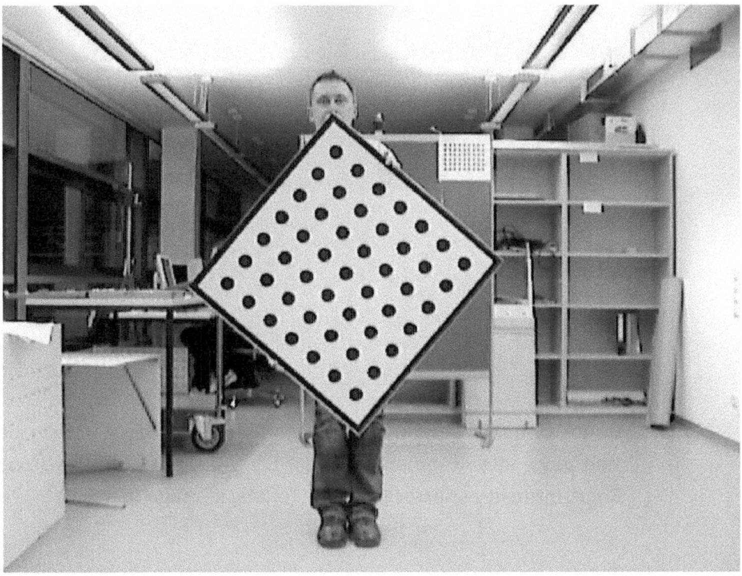

Fig. 2. Calibration plate

The above data acquisition procedure results in pairs of synchronously recorded gray-values and intensity images of the employed planar pattern. When a sufficiently large number of pairs has been acquired, the calibration procedure can be started off-line.

In order to determine whether or not the calibration plate is in fact mapped onto a specific range pixel, a simple segmentation algorithm (see e.g. [6]) can be applied to the quotient image of each pair of corresponding intensity images. The correctness of this approach is justified by Eq. (2), considering that all pixels behave in a similar way. Fig. 3 illustrates a mask image, where the highlighted range pixels of Fig. 3 (c) indicate the calibration plate.

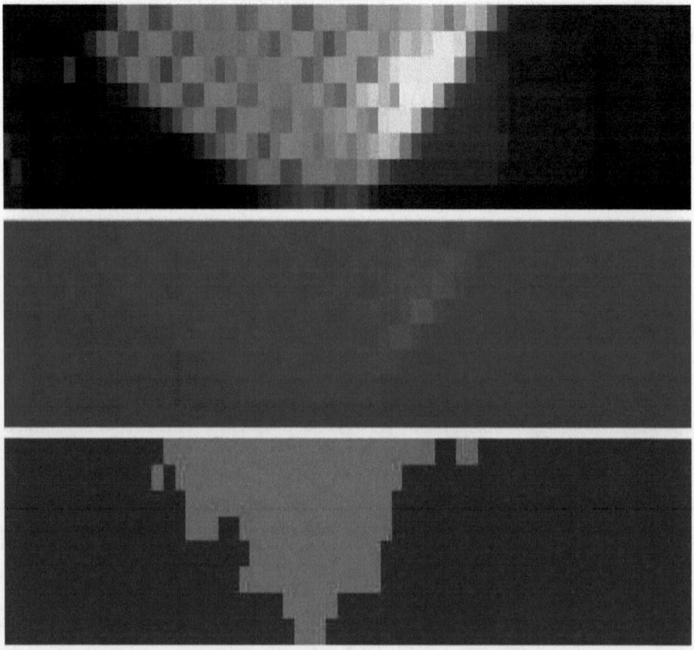

Fig. 3. Intensity images and the corresponding mask image of the calibration plate after segmentation. (a) Long intensity shutter (top), (b) Short intensity shutter (middle), (c) Mask image (bottom).

3 Calibration Procedure

The reference coordinate system is defined by the coordinate system of the gray-value camera. It is assumed that the perspective center and the optical axis of the gray-value camera coincide with the origin and the Z-axis of the reference coordinate system, respectively. It is assumed furthermore that the exposure direction corresponds to the positive direction of the Z-axis.

The calibration procedure is conducted for each range pixel separately. Suppose, for some range pixel one has obtained the measurement pairs, (s_1, l_1), …, (s_n, l_n) and, by making use of the gray-value camera, the corresponding plane equations

$$a_1x + b_1y + c_1z = d_1, \ldots, a_nx + b_ny + c_n = d_n \tag{5}$$

The employed data acquisition procedure implies that $a_i = b_i = 0$ and $c_i = 1$ for all $1 \leq i \leq n$. This simply means that the optical axis of the gray-value camera is orthogonal to each of the planes. The corresponding plane equations thus simplify to $z=d_i$; that is, all points (x_i, y_i, z_i) of the plane i have the same distance d_i to the XY plane of our reference coordinate system.

The range calibration thus yields the following least-squares problem:

$$\left\| \begin{pmatrix} Q_1 & 1 \\ \vdots & \\ Q_n & 1 \end{pmatrix} \begin{pmatrix} a \\ b \end{pmatrix} - \begin{pmatrix} d_1 \\ \vdots \\ d_n \end{pmatrix} \right\| \to \min \tag{6}$$

Here, the shutter quotients are denoted by $Q_i=s_i/l_i$. The optimization problem (6) can be solved by using a pseudo-inverse or, in an efficient and numerically stable way, using Householder transformations [7].

It remains to be shown that the optimization problem in (6) is based on correct assumptions. In our case we have to show that the linear model also applies to the Z coordinate of points in our reference coordinate system: In other words there exist some $p', q' \in \mathbf{R}$ which obey the following equation:

$$z(s_i, l_i) = d_i = p'\left(\frac{s_i}{l_i}\right) + q' \tag{7}$$

Suppose the considered pixel has the range parameters p and q as given in Eq. (2). As before, the somewhat idealized assumption is made that the distance to the plate is measured along the viewing ray $\{c+\mu r : \mu \in \mathbf{R}_0^+\}$ where $c=(c_x, c_y, c_z)$ denotes a point on the sensor surface and $r \in \mathbf{R}^3$ is normalized, *i. e.* $\|r\|=1$. Note that all coordinates are given in the reference coordinate system. With regard to the above considerations, the coordinates of the observed plane points $w_i=(x_i, y_i, z_i)$ are

$$w_i = c + d(s_i, l_i) \quad r = c + p\left(\frac{s_i}{l_i}\right) r + qr \tag{8}$$

Particularly the Z coordinate of each pixel is thus given by

$$d_i = c_z + p\left(\frac{s_i}{l_i}\right) r_z + q r_z \tag{9}$$

Setting $p'=pr_z$ and $q'=c_z+qr_z$ the following result has been shown:

The Z coordinate of points on a viewing ray of one range pixel is linear in the intensity shutter quotient s/l in any coordinate system obtained by an isometric transformation from the coordinate system of the range camera.

The validity of the optimization problem in (6) is a direct consequence of this result. For the solution of (6), neither the viewing ray nor the transformation between the coordinate systems of the gray-value and the range camera must be explicitely taken into account.

Fig. 4. UseRCams range camera prototype

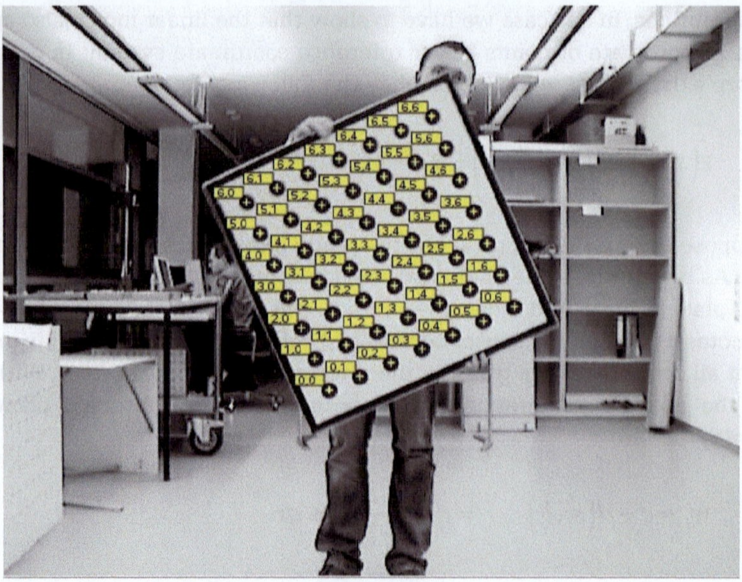

Fig. 5. Calibration plate with identified points

4 Experimental Results

Experimental results were conducted with an MDSI range camera prototype developed by the partners of the European project UseRCams led by Siemens CT. The range camera has a resolution of 64x8 range pixels. The prototype is shown in Fig. 4.

The experimental calibration plate shown in Fig. 5 has been designed to be easily recognized by the gray-value camera in a variety of distances. Fig. 5 depicts a gray-value image with an overlay of identified calibration points. The plane reconstruction algorithm used for this plate follows the ideas in [4].

Fig. 6 shows some low resolution sample intensity images of the planar calibration plate. Note that the intensity images are scaled according to the aspect ratio of the range camera pixels.

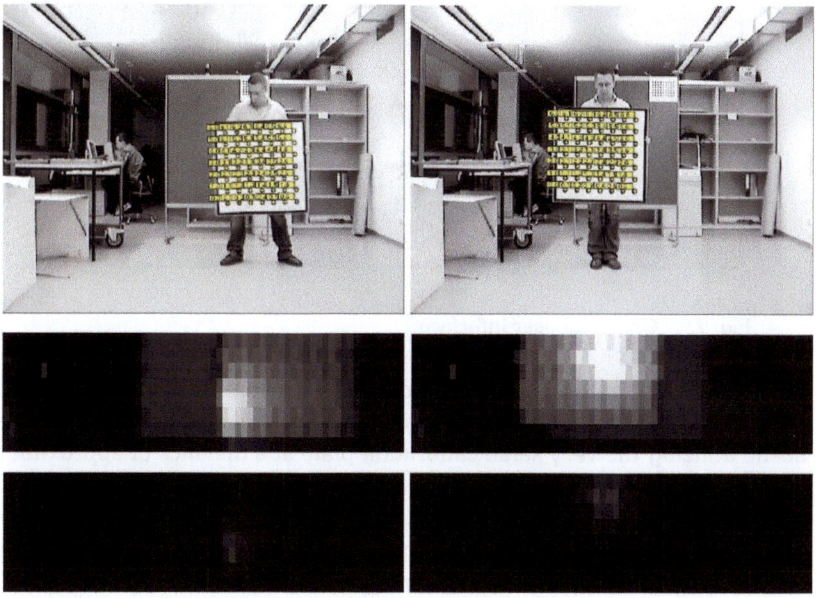

Fig. 6. Gray-value and corresponding intensity images

All algorithms described in this paper are implemented and integrated into a suitable graphical user interface. The required user interaction is negligible: It only calls for the kick-off of the acquisition process and, after acquisition of sufficiently many image pairs, for the activation of the calibration procedure. The progress of image acquisition is controlled by the underlying application: A visual feed-back signal is displayed after each range pixel has collected sufficient data for accurate calibration. Fig. 7 shows the off-line user interface.

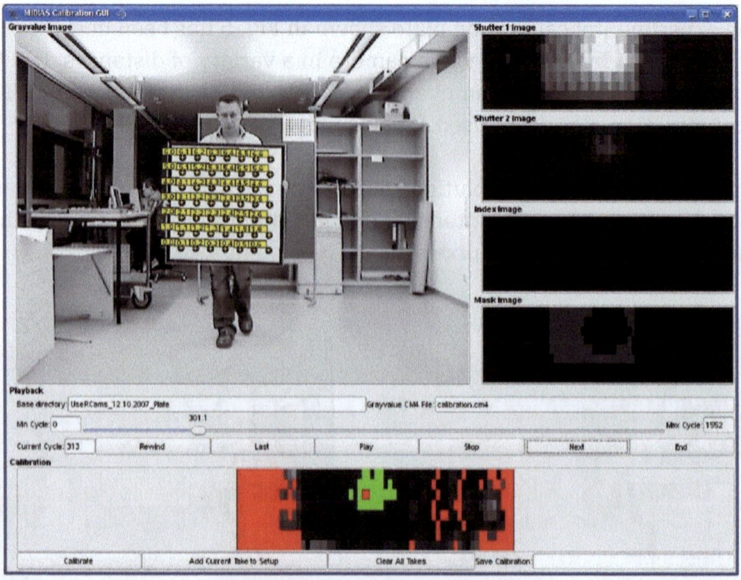

Fig. 7. Graphical user interface

The described range camera calibration procedure has been applied to an indoor environment. Subsequently, image sequences resulting from a distinct scene utilizing the very same calibration plate at a distance of approximately 5 meters were evaluated to cross-validate the primary calibration results. For the cross-validation the same data acquisition procedure was used as for the initial calibration. For each calibrated range pixel the absolute range reconstruction error $e(s_i, l_i, d_i)$ has been measured with respect to the calibration plate

$$e(s_i, l_i, d_i) = p\left(\frac{s_i}{l_i}\right) + q - d_i \tag{10}$$

To obtain comprehensive results for our experiment we have assumed the range reconstruction error of all range pixels to be identically distributed. With this assumption one can aggregate sample data from all pixels as if being taken from a single population. Fig. 8 shows the distribution of the absolute range reconstruction error for all calibrated pixels.

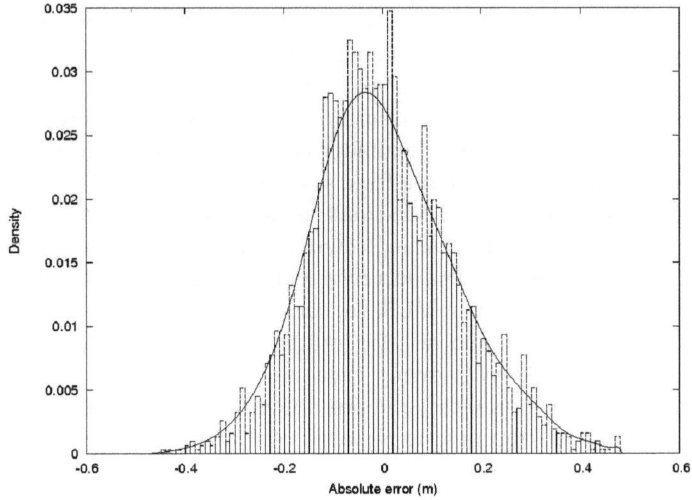

Fig. 8. Distribution of the range reconstruction error

From the distribution in Fig. 8 and the above assumptions it can be concluded that errors with an absolute value of more than 16.5 cm occur in less than 25% of all cases and that errors with an absolute value of more than 29.5 cm occur in less than 5% of all cases. The error expectation of 0.02 cm confirms the accuracy of the novel calibration procedure and the correctness of our model. In other words, the measured errors originate primarily from the sensor noise.

5 Conclusions

We have presented a flexible low-cost, and simple-to-use calibration procedure for an MDSI range camera utilizing an intrinsically fully calibrated gray-value camera. The proposed procedure takes advantage of the superior range reconstruction quality of algorithms for gray-value cameras for known objects. The price one has to pay for such flexibility is the requirement to keep the gray-value and the range camera in a fixed relation to each other once the calibration has been accomplished. For small deviations of the coordinate systems the corresponding error can be neglected considering the measurement deviations of the range camera. The accuracy and applicability of the new approach has been evaluated using an MDSI sensor prototype developed in the project UseRCams. As a challenging task for future research the complete reconstruction of the range camera viewing rays is considered; this will significantly elaborate the calibration technique presented in this paper. A fully calibrated range camera may even help to replace coordinate-measuring devices in the future.

Acknowlegdments

This work was partially supported by the project MIDIAS funded by the German Federal Ministry of Eduction and Research (BMBF) under grant no. 16SV2262.

References

[1] P. Mengel; G. Doemens; L. Listl, "Fast range imaging by CMOS sensor array through multiple double short time integration (MDSI)," in International Conference of Image Processing (ICIP), Thessaloniki, Greece, October 2001, IEEE, vol. I, pp. 169–172.

[2] O. Elkhalili; O. M. Schrey; P. Mengel; M. Petermann; W. Brockherde; B. J. Hosticka, "A 4 x 64 pixel CMOS image sensor for 3-d measurement applications.," IEEE Journal of Solid-State Circuits, vol. 39, pp. 1208–1212, 2004.

[3] Z. Zhang, "A Flexible new technique for camera calibration," Tech. Rep., Microsoft Research, 1998, Technical Report MSR-TR-98-71.

[4] R. Hartley and A. Zisserman, Multiple View Geometry in Computer Vision, chapter Epipolar Geometry and the Fundamental Matrix, Cambridge University Press, 2000.

[5] W. Förstner, "Uncertainty and projective geometry," in Handbook of Computational Geometry for Pattern Recognition, Computer Vision, Neurocomputing and Robotics, E.Bayro-Corrochano, Ed. Springer, 2004.
[6] Y.J. Zhang, "A survey on evaluation methods for image segmentation," Pattern Recognition, vol. 29, no. 8, pp. 1335 – 1346, August 1996.
[7] C.L. Lawson and R. J. Hanson, Solving Least Squares Problems, Prentice-Hall, Englewood Cliffs, NJ, 1974.

Tobias Hanning
University of Passau
Innstr. 43
94032 Passau
Germany
tobias.hanning@uni-passau.de

Aless Lasaruk
FORWISS, University of Passau
Innstr. 43
94032 Passau
Germany
lasaruk@forwiss.uni-passau.de

Reiner Wertheimer
BMW Group Forschung und Technik
Hanauer Str. 46
80937 München
Germany
reiner.wertheimer@bmw.de

Keywords: range camera, calibration, gray-value camera

Laserscanner Based Cooperative Pre-Data-Fusion

F. Ahlers, Ch. Stimming, Ibeo Automobile Sensor GmbH

Abstract

The Cooperative Pre-Data-Fusion is a novel approach for a cooperative environment perception system. It is being developed within the integrated research project SAFESPOT under the subproject SAFEPROBE which specifies and develops an in-vehicle sensing platform. This paper describes the approach of fusing Laserscanner data with information of surrounding vehicles, transferred to the host-vehicle using wireless network technology in a cooperative way.

1 Introduction

SAFESPOT is an integrated research project co-funded by the European Commission Information Society Technologies among the initiatives of the 6th Framework Program. The objective is to understand how intelligent vehicles and intelligent roads can cooperate to produce a breakthrough for road safety. Therefore, the general aim of the project is to create a Safety Margin Assistant, detecting potentially dangerous situations between road users of any kind in advance [1].

For a clear project structure, the development of the sensing platforms is separated from the application development. This yields two subprojects, one for the in-vehicle platform called SAFEPROBE and one for the infrastructure platform referred to as INFRASENS.

Although the architectures of both platforms have strong similarities and also foresee the Cooperative Pre-Data-Fusion, this paper describes and focuses on the Cooperative Pre-Data-Fusion of the in-vehicle platform developed within SAFEPROBE specifically.

The remainder of this paper is organized as follows: In the second chapter, the purpose of the Cooperative Pre-Data-Fusion is described. The third chapter explains the architecture and the full data processing chain in detail. The paper will close with a conclusion and acknowledgments.

2 Cooperative Pre-Data Fusion

The Cooperative Pre-Data Fusion is a new approach for an environment perception system using the in-vehicle SAFEPROBE multi-system platform.

Conventional automotive data fusion systems for environment perception data fuse all relevant information gathered by in-vehicle sensors only. This might be laserscanners [3] and camera systems or radar sensors.

This new approach fuses information about other road users, transferred to the host-vehicle using wireless network technology, in a cooperative way. Within SAFESPOT and therefore in the remainder of this paper, the data transferred via the wireless network is referred to as Vehicular Ad-hoc Network data (VANET-data).

The fusion of environment perception data gathered by thelLaserscanner with VANET-data is performed prior to object-level fusion with further sensor data such as radar or camera system data.

Therefore, this approach is referred to as Cooperative Pre-Data-Fusion (CPDF).

3. System and Data Fusion Architecture

3.1 Data Sources

The CPDF makes use of four data sources feeding into different levels of the pre-data fusion process as shown in Fig. 2.

- laserscanner data: range profile of the environment of the host vehicle [3]
- VANET-data: static and dynamic information transferred from one vehicle to the host vehicle using wireless network technology [5]
- Motion data of the host vehicle
- Static map data: precise digital map [4] provided by the Local Dynamic Map (LDM) running on the main PC

The output of this fusion process is a reliable and robust object description in the field of view of the laserscanner. The object description contains state estimation and classification information for basically any road user detected (e.g. passenger cars, trucks, bikes, and pedestrians).

3.2 Laserscanner Sensor

The key component of the CPDF is the laserscanner installed into the front bumper of the host vehicle. The host vehicles are conventional passenger cars as well as trucks. The processing and fusion of the data is performed on a laserscanner ECU which is connected to a VANET-router, a vehicle gateway, and a main PC providing access to a digital map.

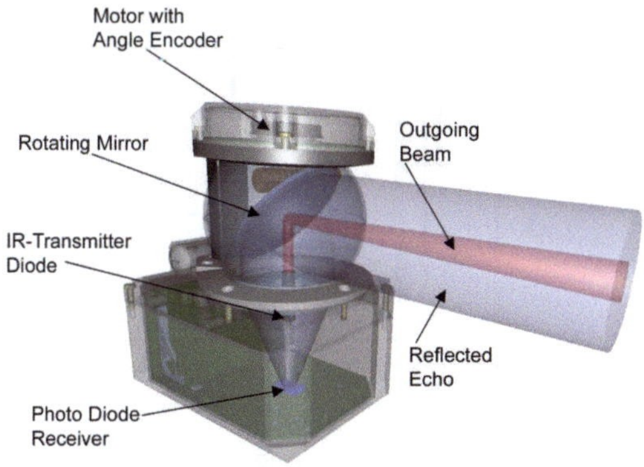

Fig. 1. Schematic of the used laserscanner

The laserscanner (shown in Fig. 1) observes its environment horizontally and gathers a range profile of the host-vehicles vicinity of up to 200 meters. The rotational frequency of the mirror guiding the laser beam is 12.5 Hz. The measurement frequency enables distance measurements every 0.25 degrees horizontally. The applied multi-echo technology ensures proper performance even under adverse weather conditions like rain, snow, or any other precipitation.

The range profile gathered during one revolution of the laser beam is referred to as one laser scan, which in turn consists of many scan points (distance measurements).

3.3 Pre-Processing

Fig. 3a shows a range profile captured by the laserscanner for the example environment scene of an intersection. The captured raw data is sent to the scan data pre-processing and segmentation sub-module of the CPDF.

The pre-processing uses its knowledge about the physical characteristics of the laserscanner and its measurement technology to categorize the range profile.

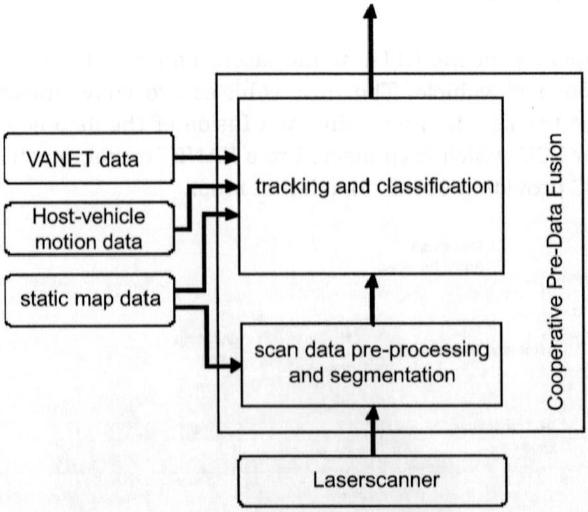

Fig. 2. Architecture overview of the Cooperative Pre-Data-Fusion

This pre-processing includes, for example, a ground detection algorithm, which marks all scan points measuring the road surface in order to exclude them from further processing. This is useful as an ideal range profile would only include scan points reflected by obstacles and road users, but not the road surface.

3.4 Data Fusion and Tracking

Another part of the pre-processing is the background elimination. This step utilizes digital map data which is taken from the highly precise LDM and the knowledge of the host vehicle's precise position calculated by the positioning system. The LDM and the positioning system are both developed within a third SAFESPOT sub-project called SINTECH. Both are part of the onboard SAFESPOT system and therefore always available to the CPDF.

In order to distinguish between background objects (e.g. bushes, buildings, and front yards) and foreground objects (e.g. passenger cars, trucks, bikes, and pedestrians), the surrounding static road and lane geometry is compared to the range profile (see Fig. 3b). All scan points of the current range profile outside the boundaries of the surrounding road, sidewalk, and traffic islands are marked as background and not further processed (Fig. 3c). During this elimination process, it is important to take accuracy estimations of the host vehicle's position and orientation into account as well as accuracy estimations of the digital map. This is needed in order to avoid false elimination of foreground objects. With this background elimination, the overall processing performance and robustness is increased significantly.

Subsequently, a segmentation algorithm is performed that groups single scan points into clusters based on a weighted distance between two scan points. Ideally, one cluster represents one real object such as a passenger car or pedestrian. These clusters are forwarded to the tracking and classification module. Such clusters are called segments.

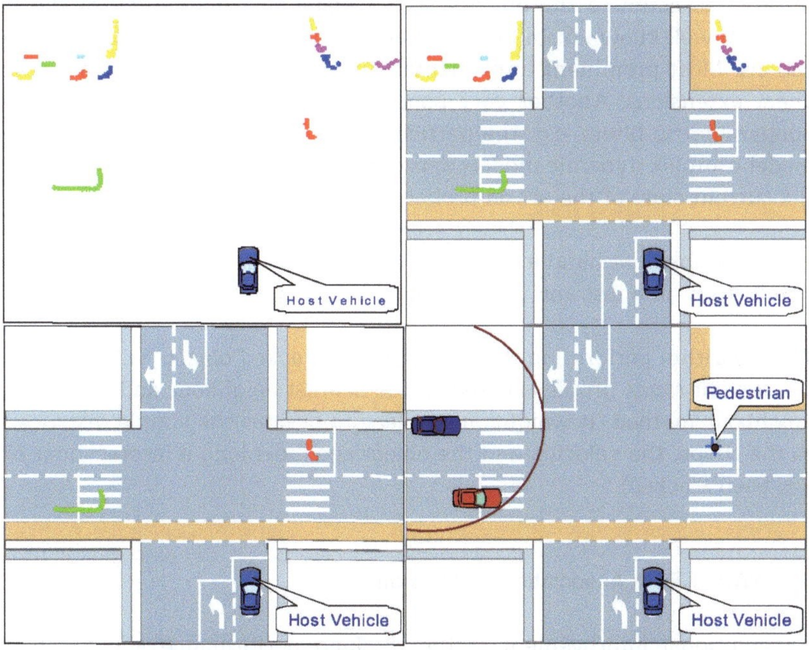

Fig. 3. Steps of the Cooperative Pre-Data-Fusion showing (a) a raw range profile, (b) map association, (c) tracked objects, and (d) cooperative VANET-data association.

3.5 Laserscanner Object Tracking

The tracking of established laserscanner objects is performed by comparing one segment's parameters with predicted parameters of known objects from the previous scans. Such parameters are the reference point coordinates of the segment and object. The reference point is the centre of gravity of all scan points belonging to the segment in the most simple case (e.g. for pedestrians). Under certain conditions, the segment's geometry is analysed and geometrical reference points based on the segments shape are used. This is the case for passenger cars, trucks, and busses, already classified. The benefit of this alternative method is a more precise position, velocity, and course angle estimation.

Segments that do not correspond to any object's parameters of the previous scans are instantiated as new objects, initialised with default dynamic parameters.

In order to estimate the object's state parameters, derivatives of the Bayes filter are well known from literature and applied successfully in various applications. Most prominent implementations are the Kalman filter and its nonlinear extensions. Another very popular branch of implementations are the nonparametric filters, e.g. the particle filter. With these filters one is able to model complex dynamic models appropriately. An extensive introduction into the current state of the art on probabilistic filter algorithms is given in [2].

For the problem at hand a Kalman filter approach is chosen. Our implementation focuses on a fast and precise object tracking.

The algorithm performing the association of tracked objects with segments of the current range profile is based on the nearest neighbour and global nearest neighbour method. It weights many factors besides the simple distance such as the shape, the velocity, and the object age (how long a laserscanner object has been tracked).

3.5 VANET Data Association and Fusion

An additional improvement of the tracking performance and reliability is achieved by fusing static and dynamic vehicle information transferred via wireless vehicle-to-vehicle communication into the filter.

The major challenge within this cooperative approach is an accurate and robust association of VANET-data to the correct vehicle detected by the laserscanner. In case, the transmitting vehicle's position is based on SINTECH's positioning system, its position accuracy will be fairly high and therefore easy to associate to the correct laserscanner object.

In a more complex but also likely case, the transmitting vehicle is not equipped with such a positioning system. Most likely, the position estimation would be based on a standard GPS signal and therefore be quite inaccurate. As an example, one could assume that the position estimation based on GPS has a circle of 95% probability with a radius of 15 m (see Fig. 3d). This would result in many association candidates for dense traffic scenarios and thus make a correct association based on the position estimates very demanding. Consequently, further parameters are required to find the correct laserscanner object. Most obvious parameters are the object's velocity and course angle, its size (length and width) and also its type (e.g. passenger car, truck, bus, motor bike). Based on this additional static and dynamic information on the transmitting vehicle and the knowledge on the laserscanner object, an association likelihood is calculated for each object. The most likely object is then associated and the object parameters are fused.

Fig. 3d shows a scene where the red vehicle sends its information about itself to the host-vehicle. The transferred position is depicted as a red striped vehicle while the red circle illustrates the circle of 95% probability of the vehicle's true position. In case of an unblocked view and only a few road users, the association is straight forward. However, in dense traffic scenarios more complex analysis and association technique as described above is required.

3.6 Object Classification

The final step of the whole Cooperative Pre-Data Fusion process is the object or road user classification. This process determines the road user's type:

- passenger car,
- truck or bus,
- bike,
- pedestrian, or
- unknown object (default).

The likelihood for each class is calculated by the classification. Currently, it is state of the art to associate the object to one class by comparing its geometric information like its shape and dynamic data such as its velocity to typical representatives of each class.

One optimization is achieved by adapting the probability for an object class based on its position within the static map. One example would be an object on the sidewalk. Its likelihood of being a car would decrease while the likelihood of being a pedestrian would increase. In SAFEPROBE, the procedure is done automatically by feeding all this information into a specialized, well trained artificial neural network (ANN) [5].

On top of this optimization, additional information on the object provided via VANET is fused into the classification process [6]. By performing this fusion, it is assumed that the vehicle's knowledge on its own class is very high and is therefore fused using a high probability value.

After the Cooperative Pre-Data Fusion process is finished, the gathered object information data is sent to the object-level fusion. It fuses and processes the object information collected by all installed environment perception systems specified within SAFEPROBE.

The subsequent time step of the CPDF is triggered by the next scan sent to the laserscanner ECU. The full process starts over again using the road user's position and classification information as optimized through the cooperative fusion.

4 Project Progress

The described system is currently under investigation. A SAFEPROBE prototype system is being built and real-life performance tests are being prepared. So far, the first simulations and tests indicate quite promising results, but quantitative measurements will not be available before mid-2008.

5 Conclusion

As described within the previous chapters, the Cooperative Pre-Data-Fusion increases the reliability and robustness of standard tracking and classification based on laserscanner data only. It therefore enables many safety relevant applications developed within SAFESPOT.

Acknowledgement

SAFESPOT is part of the 6th Framework Programme, funded by the European Commission. The partners of SAFEPROBE and INFRASENS thank the European Commission for supporting the work of this project. The authors would like to thank all partners within SAFESPOT especially from SAFEPROBE, INFRASENS and SINTECH for their cooperation and fruitful teamwork on data fusion and data processing.

References

[1] SAFESPOT - Co-operative Systems for Road Safety 'Smart Vehicles on Smart Roads'; online at: www.safespot-eu.org last checked: 03.01.2008.

[2] S. Thrun, W. Burgard, and D. Fox, Probabilistic Robotics, MIT Press, Cambridge, MA, 2005.

[3] K. Fuerstenberg and K. Dietmayer, Object Tracking and Classification for Multiple Active Safety and Comfort Applications using a Multilayer Laserscanner. In: Proc. IEEE Intelligent Vehicle Symposium IV04, Parmy, Italy, June 2004.

[4] M. Garrelts, M. Mittaz, A. Varchmin, W. Vogt, Navigation-based driver assistance systems. ITS European Congress 2005, Hannover, Germany

[5] M.T. Hagan, H.B. Demuth, and M.H. Beale, Neural Network Design, Boston, MA, USA: PWS Publishing, 1996.

[6] S. Wender, T. Weiss, K. Dietmayer, Integration of Wireless Communication Data in a Sensor Based Vehicle Environment Model, In: Proc. 4th International Workshop on Intelligent Transportation, Hamburg, Germany, March 2007.

Florian Ahlers, Christian Stimming
Ibeo Automobile Sensor
22143 Hamburg
Germany
florian.ahlers@ ibeo-as.com
christian.stimming@ ibeo-as.com

Keywords: laserscanner, cooperative safety, vehicular ad-hoc networks, data fusion, safety marging assistant, environment perception

Satellite-Based System for Predictive Control in Vehicles and its Field of Applications

A. Zlocki, Institut für Kraftfahrwesen Aachen

Abstract

A satellite-based system for predictive control enhances advanced driver assistance systems in the fields of comfort, safety and environment. The predictive information of the GNSS-receiver in combination with a next generation digital map are fused with local environment information of the vehicle's internal sensors and a distance sensor in the front of the vehicle. In comparison to conventional ADAS sensors, which can only take the distance of 200 m in front of the vehicle into account, ADAS with predictive sensor information can react to driving events such as velocity limitations, curve radii, intersections, road inclinations and traffic signs in an early stage and therefore help to increase the driving comfort and traffic safety and increase the traffic efficiency and reduce fuel consumption as well as CO_2 emissions. The development of the predictive preview system as well as the integration into test vehicles have been conducted at the ika (Institut für Kraftfahrwesen Aachen). Two different applications are presented as examples and the benefit of the satellite-based system is described in simulation and test results.

1 Introduction

In the past decade technical systems have been developed in order to support the driver in complex driving situations and relieve him of monotonous driving tasks, due to the increasing traffic volume and the resulting complexity. These systems are so-called advanced driver assistant systems (ADAS). ADAS help to compensate the known weakness of a human driver (inattentiveness, reaction times etc.) and therefore are promising to decrease accident numbers.

ADAS are interlinked with all three elements of the closed control loop "traffic" (vehicle, environment and driver) as given in Fig. 1. Additionally there are systems available, which partly take over the driving task and replace the driver in defined tasks completely, e.g. ESP, Lane Keeping and ACC (adaptive cruise control). In these cases the control loop consists of the vehicle, the environment and the advanced driver assistance system.

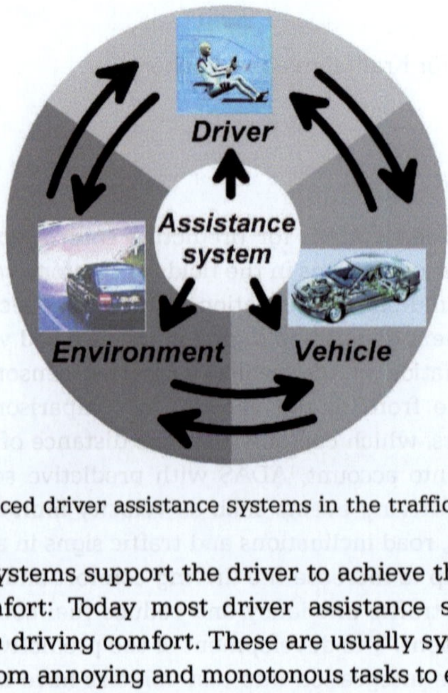

Fig. 1. Advanced driver assistance systems in the traffic system

Driver assistance systems support the driver to achieve the following aims:

- Driving comfort: Today most driver assistance systems concentrate especially on driving comfort. These are usually systems, which relieve the driver from annoying and monotonous tasks to ease the drive. Those systems can also have an indirect positive impact on traffic safety.
- Safety: Advanced driver assistance systems can be used to prevent accidents, to decrease accident damages and improve the rescue management. Therefore, these driver assistance systems either completely take over the vehicle control or give additional information or warning to the driver.
- Traffic efficiency: An improvement of street capacity is expected by the usage of driver assistance systems. Thus traffic jams can be prevented or dissolved faster. In addition the vehicles, which are approaching the traffic jam, can be redirected automatically.
- Environment: The support of the driver in form of technical systems can be used to reduce fuel consumption as well as CO_2 and noise emissions. The assistance system can give a suggestion for a driving style depending on the situation and the foresight of the traffic in front. For complex drive train structures (hybrid engine) the operating strategy is set to an optimum.

The driver controls the vehicle in longitudinal and lateral direction. At the same time he has to cover various driving tasks, which can be classified into three main categories using the three level model shown in Fig. 2 [1]. These three levels are:

- navigation
- guidance
- and stabilisation.

ADAS are able to support the driver in each of the three levels. Today sophisticated systems provide assistance in even more than one level only. Depending on the assistance, the vehicle needs to be equipped with sensors and actuators. The control algorithm determines the desired set values and operates actuators as for example throttle position, steering wheel angle or adjustment of the headlights' range.

Depending on the ADAS application the sensor systems have to detect the surrounding environment. Local traffic conditions are detected by distance sensors based on radar or lidar technology with a distance of up to 200 m in front of the vehicle. Image processing systems detect objects on the driving route, lane markings or recognise traffic signs. An extended field of preview with a range of more than 200 m independent of curve radii and direct field of view allows the introduction of new applications.

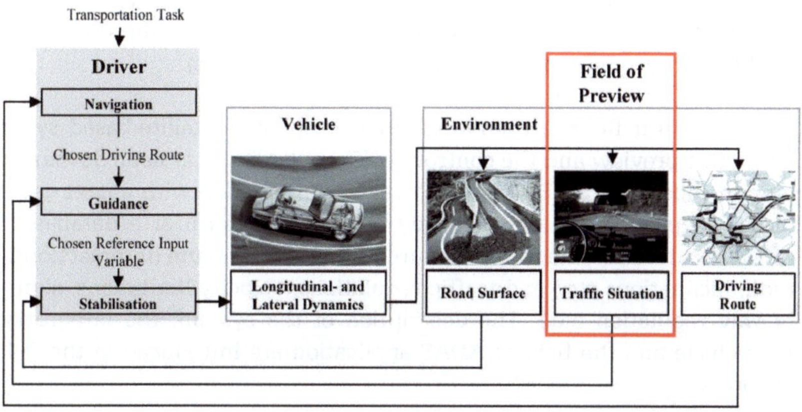

Fig. 2. Three level model and the closed control loop vehicle guidance

The determination of the position on the driving route is conducted by means of a global navigation satellite system (GNSS). Navigation systems, which support the driver on the navigation level, are already widely popular, as modern receiver units can fit easily into small pocket pc or mobile phones based on their small cross-section and low prices.

Today two GNSS-systems are available, the American NAVSTAR Global Positioning System (GPS) and the Russian GLObal NAvigation Satellite System (GLONASS). GALILEO, the European systems, is currently under construction.

Since the 2nd May 2000 the GPS-system provides accuracy below the range of 10 m depending on the number of satellite signals, which are received at the same time. Differential GPS increases the accuracy by transmitting a correction signal by radio communication from a ground station at a precise known position. GALILEO will provide different services with higher accuracies. Accuracies within the meter range (< 4 m in the open service and < 1 m in the commercial service) and a reliable detection of problems within 10 seconds as well as a maximum coverage are the expected advantages of this civil satellite navigation system.

In combination with a digital map the satellite navigation system is able to enhance the horizon of the traffic situation from 200 m up to several kilometres ahead of the vehicle. Additionally new attributes, which are impossible to detect by sensor systems are included into the digital map. This preview system is used as an additional sensor for ADAS applications. European manufacturer, supplier and research institutions are aware of the advantages of these systems and formed research programs such as NextMAP (2000 – 2001) and the PReVENT subproject MAPS&ADAS (since 2004). Japanese manufactures apply such systems since 1998 (AISIN AW's Navimatic).

At the Institut für Kraftfahrwesen Aachen (ika) a satellite-based system for predictive preview and the control of different ADAS has been developed. The system consists of a GPS/TMC-receiver in combination with a next generation digital map, which is enhanced by attributes in form of a database. These attributes describe the infrastructure of the driving route (intersections, curve radii, inclinations etc.) and traffic regulations (velocity limitations, signs, right of way regulation etc.). The description of the system, the integration into the vehicle and the field of ADAS application are introduced in the following chapters.

2 System Description

The main components of the predictive preview systems are the GNSS-receiver and the next generation digital map. These are combined with a distance sensor and information about the vehicle state as for example velocity, acceleration and yaw rate. The GNSS-receiver determines the current position of the vehicle. Additional attributes, so-called route data, are integrated into the digital map or can be added manually depending on the application of the system. This information about the driving route and the route data is used for a situation classifier in combination with local sensor data from the vehicle's sensors as well as the distance sensor. The classified driving situation and the next upcoming event is available in form of CAN-data for the controller of the ADAS application. The available number of upcoming events can be adjusted depending on the application. In order to minimise the data on the CAN-bus the number of events should be limited to a maximum of approximately 20. This provides an average preview horizon of 2 to 5 minutes for an average drive.

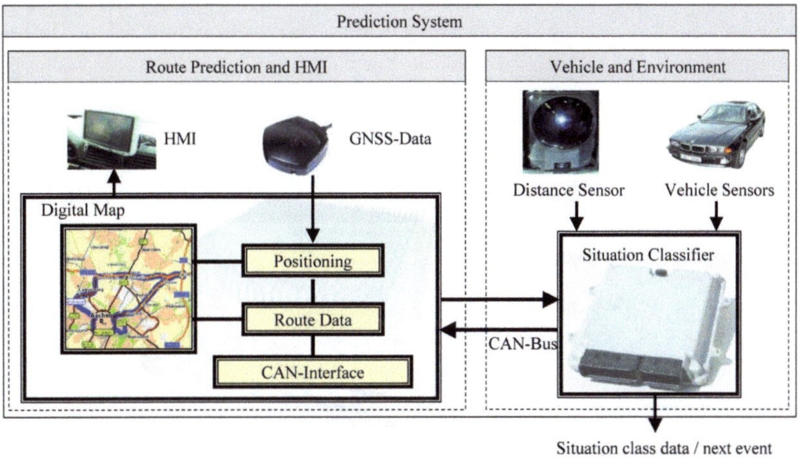

Fig. 3. Scheme of route prediction system

The software structure of the overall system is given in Fig. 4. The GNSS-receiver, the distance sensor and the vehicle sensors provide input data for the system. Traffic jam detection is realised by using a RDS/TMC (Radio Data System/Traffic Message Channel) receiver.

The vehicle`s position has to be constantly available and updated for a correct calculation of the route prediction. During tunnel situations and temporarily drop-outs of the satellite signal, because of occlusion in alley roads or high buildings at the roadside, the signal is checked for validity first. In case the signal does not provide valid data, the current position is determined by means of interpolation of previous coordinates. In case the new position data is valid a road matching is performed in which the deviation of the determined position and the available infrastructure of the digital map is minimised.

Currently available digital maps are not yet fully equipped with all necessary information. The attribute database is integrated as an additional information source. For each vehicle position the next relevant events are provided by this database on attributed routes (see section 2.2 for details).

The distance sensor detects vehicles in front of the current driving position up to 200 m on straight tracks. Out of all detected objects the relevant target is identified and handed over to the situation classification module. Statements about the local traffic condition can be derived from this sensor.

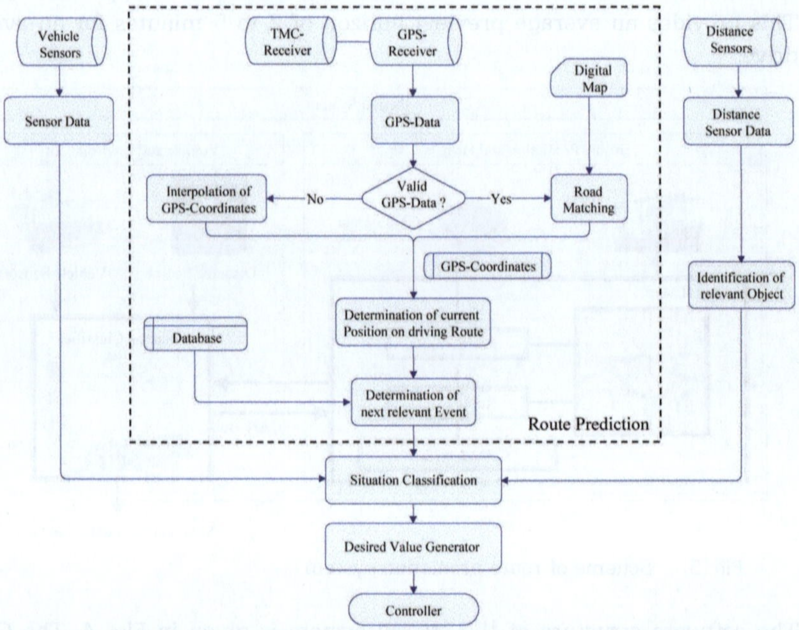

Fig. 4. Flow chart of the route prediction system

The situation classifier processes all input information and determines the current situation, which is handed over to the ADAS controller. The overall driving route is divided into incremental sections, which are in between events on the route.

2.1 Human-Machine Interface

The human-machine interface (HMI) should allow the driver to operate the ADAS and provide information about the driving route and the current status. The layout of the HMI is adapted to each ADAS application. Basically the layout is divided into status information on the left side and the visualisation of the current route on the right side, as given in Fig. 5 in two examples for the

HMI of two different systems. Therefore the HMI can also be used as a navigation system in the next development step.

The system constantly checks the status of the GNSS-receiver and the communication to the vehicle's CAN-bus. The current velocity limitation, which is calculated on basis of curve radii, signs or traffic regulations, is displayed. The next event, the distance to the next event and the value of the next event, e.g. velocity limitations, inclinations or other attributes, are shown. TMC information is also provided, if available for the chosen driving route. Optional the current situation class and the sensor data of the connected sensors are shown for control purposes.

Fig. 5a. Layout of HMI for hybrid vehicle applications.

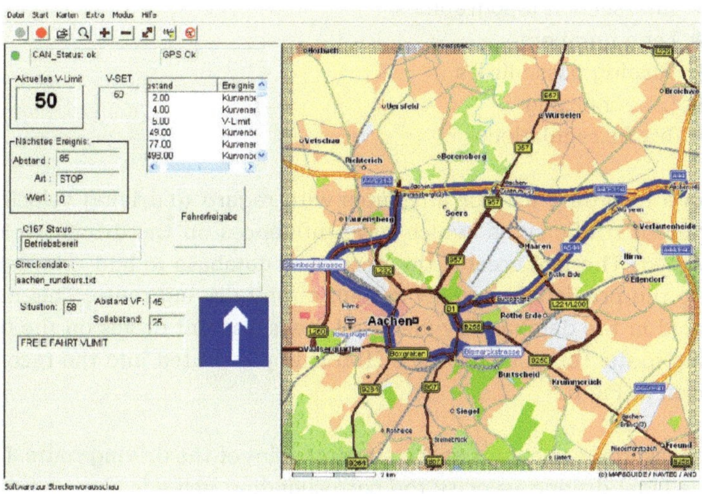

Fig. 5b. Layout of enhanced ACC applications.

2.2 Map Attributes

Today's European navigation systems provide an accurate mapping of the road infrastructure. Manufacturers of digital maps are currently working on the next generation standards. Those maps will include attributes like the number of driving lanes, inclinations and declinations, velocity limitations and curve radii. Prototype systems of regional maps are available and used within research projects.

A useful classification of map attributes is given in [4]. The two main categories are range markings, which provide the start and the end of an event within a certain range, and route markings, which provide values of events at certain positions on the driving route.

Range markings are:

- Beginning, end and value of velocity limitations
- Beginning, end and type of passing prohibition
- Beginning, end of construction sites
- Beginning, end of traffic jam areas

Position markings are:

- Curve radii
- Positions of lane markings
- Inclinations/declinations
- Positions of traffic lights
- Positions of traffic signs
- Intersection topographies
- Lane numbers
- Turning possibilities
- Lane merging
- Traffic regulations

At ika a recording tool for attributes with regard to driving velocity and the position of traffic signs, which have influences on the driving velocity, was developed. The tool creates the necessary database to add relevant information to the predictive preview system of any driving route. This information is stored in the database by means of measurement drives on the designated driving route. For this purpose a HMI is implemented into the recording tool, as given in Fig 6.

The measurement begins at the starting point of the driving route. Depending on the legal driving velocity the corresponding value is chosen from a list of icons (see Fig. 6). Traffic signs, which have relevance for the driving velocity, are also available. Curve radii have to be obtained by using the GPS-data of the measurement drive. In an offline process the curve radii are calculated and included into the database. After the measurement is conducted and the calculation of curve radii is processed the database is stored in combination with the driving route.

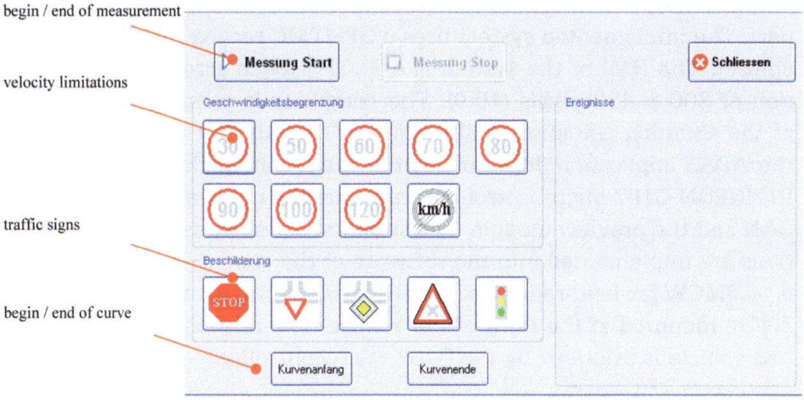

Fig. 6. HMI for manually measurement of driving routes

The list of recorded attributes can be extended depending on the ADAS application.

TMC-data are broadcasted by radio stations in most European countries with approximately 60 bit/s. Each TMC-message contains an event code and a position code to describe local traffic situations. These attributes are filtered by the predictive system for messages, which have influence on the driving velocity on the route such as traffic jams, accidents and redirections.

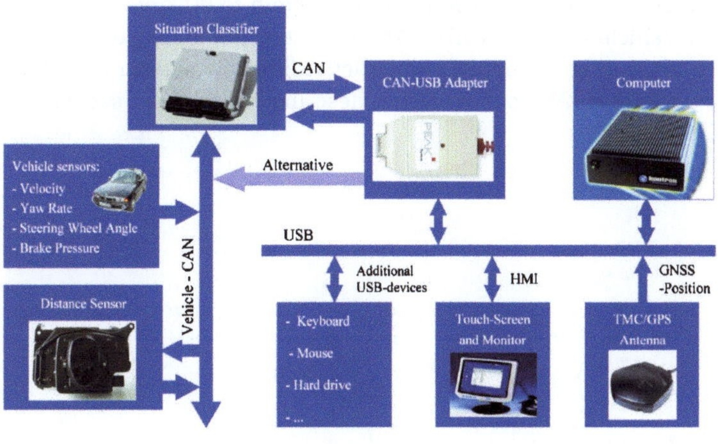

Fig. 7. Hardware components integrated into the test vehicle

3.3 Hardware Implementation and Testing

The satellite-based system for predictive control is implemented into a test vehicle at ika. The target platform of the digital map is a MS Windows XP based PC for automotive applications, which can be operated in a wide temperature range. The GNNS-receiver is connected to the computer via the USB-port. The implemented system uses a GPS/TMC-receiver with a SiRFStarIIe/LP chipset. The HMI of the system is a 7" TFT touch screen display at a resolution of 800 x 480 pixels (16:9). The computer is connected to the CAN-bus of the situation classifier or alternatively directly to the vehicle-CAN in case the ADAS application does not require the situation classifier (see Fig. 7). An INFINEON C167 microcontroller represents the gateway between the vehicle-CAN and the preview module. The situation classifier as well as ADAS applications are implemented into the software of the controller. The distance sensor is a FMCW-far-field-radar on 76 GHz basis with a range of approximately 200 m mounted at the front of the vehicle. The relevant object in front of the test vehicle is priorised by the C167 microcontroller.

The power supply of the components is provided by the ika test vehicle. Additionally a CAN- and USB-bus are available.

Test results of the implemented system are given in Fig. 8. The test vehicle was driving on the A544 motorway leaving in the direction Europaplatz in the city of Aachen. The figure shows the velocity limitations of the digital map, the distance to the next event (velocity limitations on the driving route) calculated on basis of the GPS-position and the distance to the target vehicle from the radar sensor of the test vehicle. Due to the update rate of the GPS-satellites the distance to the next event is calculated with a frequency of 1 Hz. The velocity limitation changes from 100 km/h to 80 km/h to 60 km/h. The distance to the target vehicle decreases from 140 m at the beginning of the measurement down to 30 m at the end. The predictive preview system provides the necessary data for ADAS applications as described in the next section.

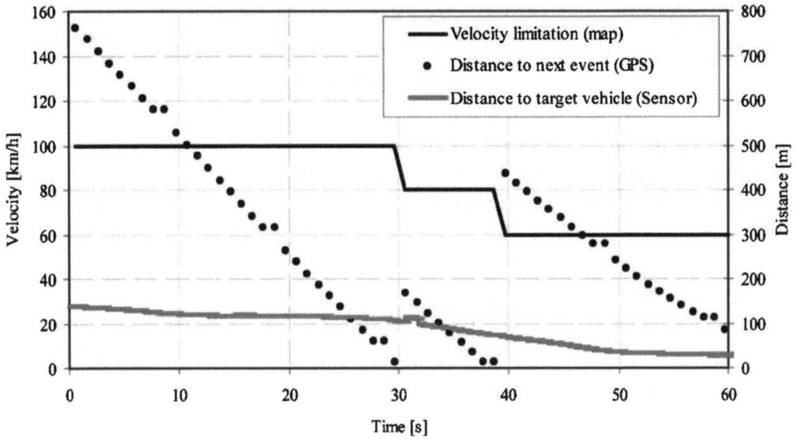

Fig. 8. Test drive result on the A544 motorway in direction of Europaplatz, Aachen

4 Field of Applications for ADAS with Predictive Preview

The field of application ranges from comfort systems, which support the driver in monotonous driving tasks and enhances functionalities of available ADAS systems to safety systems, which can warn the driver in an early stage about upcoming hazards or take over the driving task in critical situations. Furthermore, the predictive preview can increase traffic efficiency. Especially for hybrid vehicles the system increases the constancy of the driving velocity and therefore reduce fuel consumption and emissions.

The following table shows an overview of ADAS systems, which utilise the predictive preview as an additional sensor. The systems are classified into the categories safety, comfort and environment (fuel consumption and traffic volume).

Application	Function	Possible application of predictive preview
	Safety applications	
Collision Warning / Collision Mitigation / Collision Avoidance	*Warning or active intervention into the brake system before possible collisions*	*Support of distance sensors for object tracking* *Situation classification* *Prediction of direction of the host vehicle*
Adaptive Light Control	*Situation adaptive light system*	*Information about road type* *Inclination/declination* *Curve information*
Curve Velocity Warning	*Warn driver of upcoming narrow curves*	*Information about curve radii*
Curve Velocity Control	*Reducing vehicle velocity*	*Prediction of upcoming curve radii*
Lane Keeping Assistant / Lane Change Assistant	*Warning / informing or active support of driver for lane keeping / lane changes*	*Support of image processing system in complex situations, e.g. motorway junctions*
	Comfort applications	
ACC, ACC Stop&Go	*Cruise control with distance detection to target vehicle*	*Support of distance for object tracking* *Velocity limitation based on curves and traffic regulations (signs, intersections etc.)*
Autonomous Driving	*Fully automated driving*	*Precise positioning of host vehicle* *Support of environment sensors* *Static route information*
	Environment	
ISA (Intelligent Speed Adaptation)	*Adjust driving speed to certain values*	*Prevention of traffic jams by decreasing the current driving speed on crowded motorways*
Fuel Consumption Optimisation / (Hybrid) Power train Management	*Predictive drive train control / strategy*	*Information about road topology*

Fig. 9. Overview of applications with predictive preview [4]

In the following two examples of predictive preview applications are described in more detail.

4.1 Predictive Preview for ACC

The predictive preview system increases the field of application of modern ACC-systems. Not only motorway situations, but also country roads and urban roads are taken into account. Furthermore the controller behaviour of the ACC is adapted according to the corresponding driving situation. Different controller behaviour for motorways, country and urban roads are the results. Furthermore, information about surrounding road infrastructure improves driving safety, as the system automatically adapts to the velocity limits given by the legislator. This limit represents in most cases the safest and most appropriate maximal velocity for the considered road section. A more realistic adaptation to the actual driver behaviour, e.g. in stop-and-go traffic on urban and suburban roads, is possible.

Different driving situations have been derived from proband tests as described in [3]. The implemented system considers 17 different situations in total. These situations are: manual mode, standing still (urban roads), stopping on urban roads (vehicle ahead), approaching on urban roads, stopping on urban roads (infrastructure), driving off, following (urban roads), free driving considering velocity limits (urban roads), free driving (urban roads), standing still (traffic jam), stopping (traffic jam), approaching (motorway), driving off in platoon (motorway), following (motorway), free driving considering velocity limits (motorway) and free driving (motorway).

The main criteria for the difference between situations on rural roads and motorways is the legally allowed driving velocity. This is an attribute of the digital map. Based on this information the classifier differentiates between motorways with mostly high velocities (except for traffic jams) and the urban roads with low range velocities and stop and go situations (for example at traffic lights).

Furthermore, the situation classes are separated in situations with and without a target vehicle. This is of importance in order to adapt the control behaviour to the situation. Driving behind a target vehicle is much more critical than free driving or free driving considering velocity limitations based on the infrastructure. Therefore, the longitudinal vehicle control behind a target vehicle has a higher priority, e.g. the end of a velocity limitation (as an information of the digital map) has no influence on the control in case the target vehicle is remaining in a lower velocity range.

In situations with or without a target vehicle it is separated between an acceleration manoeuvre and a deceleration manoeuvre. In case a standing vehicle in the own driving lane is detected by the distance sensor, a situation class is chosen by the classifier, which describes a braking procedure such as approaching or stopping. Braking manoeuvres can become relevant without having a target vehicle in front, e.g. because of a traffic sign. The situation class "stopping on infrastructure" describes these manoeuvres, in which a controlled and smooth brake down of the vehicle is realized. This class is not available for motorway situations as stopping manoeuvres on the motorway are normally caused by the traffic ahead. Situation classes, which result in a braking of the host vehicle, are "approaching", "stopping", standing still" and "stopping on infrastructure". Situation classes such as "driving off considering velocity limitations", "driving off" as well as "following" result in acceleration.

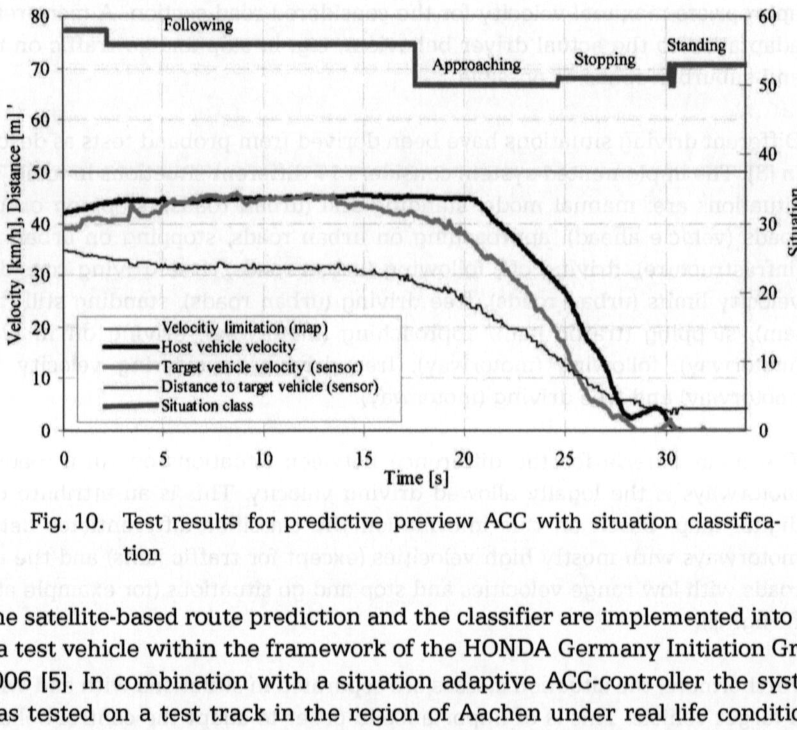

Fig. 10. Test results for predictive preview ACC with situation classification

The satellite-based route prediction and the classifier are implemented into the ika test vehicle within the framework of the HONDA Germany Initiation Grant 2006 [5]. In combination with a situation adaptive ACC-controller the system was tested on a test track in the region of Aachen under real life conditions. The test track consists of a motorway with high probability of traffic jam and stop&go traffic, urban roads and inner city driving. An exemplary test result for a driving situation in urban traffic is given in Fig. 10. The ika test vehicle is following a target vehicle at approximately 40 km/h at the beginning of the measurement. The target vehicle slows down at traffic lights and comes to a standstill. During this manoeuvre different situations classes such as free driving, following, approaching, stopping and standing still are detected by the system. The velocity limitation from the digital map is given with 50 km/h. Based on the detected situation class and the maximum velocity limitation the controller behaviour is adapted to urban roads in comparison to motorway driving.

3.3 Predictive Preview for Hybrid Drive Trains

Hybrid vehicles are currently being successfully introduced into the market. The acceptance of the combined combustion engine and electric motor drive train is a step towards environment friendly vehicles. This is widely accepted and demanded by the public and the legislator.

The operating mode strategy for hybrid vehicles is depending on various factors such as the state of charge of the battery, the power demand in the current driving situation and the driving style of the driver. The intelligence of the operating mode strategy is increased depending on the available information of the driving route. The satellite-based system for predictive control allows the improvement of this strategy as inclinations and declinations of the road ahead as well as curve radii and velocity limitations are known. Therefore the necessary power demand is estimated and integrated into the vehicle control. Further improvement of the operating mode strategy is achieved by partly taking the driver out of the loop and automating the longitudinal vehicle control by means of ACC.

In a first step software in the loop (SIL) simulations with the traffic simulation tool PELOPS [2] have been conducted. PELOPS represents a combination of vehicle-, environment and traffic specific models. Its advantage is the possibility to take into consideration all interactions occurring between driver, vehicle and traffic. The core of the program is formed by the three essential elements: traffic systems stretch/environment, driver and vehicle. These elements enable PELOPS to simulate traffic in accuracy unequaled so far.

In order to simulate a complex drive train and energy management of a hybrid vehicle the vehicle model and the operating mode controller is designed in MATLAB/SIMULINK. This model represents the longitudinal dynamics of the vehicle including all driving resistances and engine characteristics. This vehicle model is connected to the driver and environment model of PELOPS. The predictive preview algorithm is implemented into PELOPS and can therefore substitute the digital map and the GNSS-system with PELOPS data of the environment module.

In order to decrease the fuel consumption furthermore an ACC-controller is implemented into the vehicle model in MATLAB/SIMULINK. Instead of the driver the ACC-controller provides the acceleration demand to PELOPS.

PELOPS calculates the preview data, which are handed over to the vehicle module, based on the program's environment data.

Simulation results of the impact of the route prediction and the ACC-controller for a test scenario are given in Fig. 11. The diagram shows the fuel consumption of the vehicle model in a sub-urban scenario. As a reference a PELOPS driver is compared to different parameter settings of the ACC-controller (time gaps, controller characteristics, point of acceleration/deceleration before/after events).

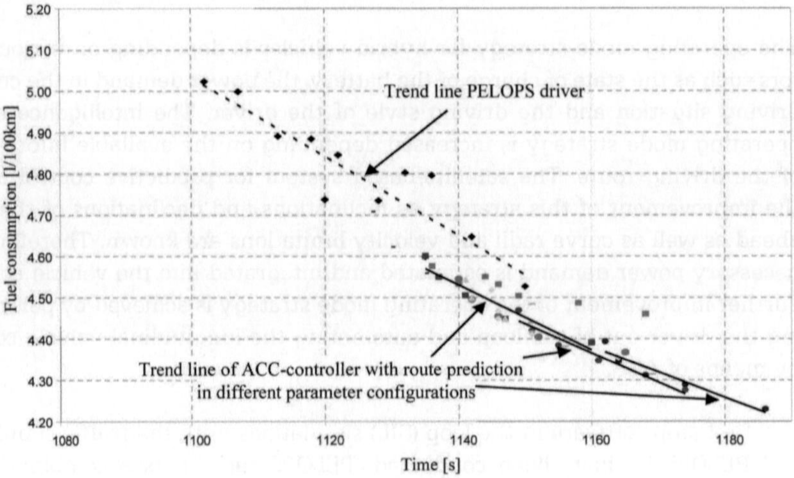

Fig. 11. Fuel consumption with route prediction and ACC-controller [7]

The fuel consumption is dependent on the driving time. Different PELOPS driver provide variations in the time/fuel consumption ratio depending on their driving behaviour and driving style. The ACC-controller with predictive control provides a fuel consumption reduction of 0,8 l/100 km in comparison to the driver model of PELOPS without preview. This is a reduction of about 20 %. In the slowest case the driving time increases from 1100 s to 1185 s, which is an increase of 7 %. For an equal driving time (different ACC-parameters) the fuel consumption reduction provides 3 % in comparison to the PELOPS driver [6].

The evaluation of the system on a test track in the region of Aachen is conducted within the framework of a BMBF supported project. The target of the project is to build up a hybrid vehicle with route prediction and maximize the fuel consumption reduction [7].

5 Summary and Outlook

Future ADAS applications will be equipped with an electronic horizon. Digital maps will provide attributes about the driving route, which provide much more information than the road network of today's digital maps for navigation systems. The civil European satellite navigation system GALILEO is expected to provide the vehicle position within accuracy in the meter range.

A satellite-based system for predictive control of vehicles is developed and implemented in a test vehicle at the ika. Several test routes in the region of Aachen are recorded and additional attributes such as curve radii, velocity limitations, traffic signs and traffic regulations are recorded and implemented into the database of the test route. The predictive control system provides the basis for various ADAS applications. The traffic status of parts of the motorway in the region of Aachen is available by means of TMC-data.

The functionality of the system is shown for two different ADAS applications. A description of the systems and first test results of the prototypes are given. Additional applications for the predictive system are mentioned.

The system works on motorway, sub-urban and urban roads. Within inner city limits high buildings prevent the reception of satellites signals from the necessary number of satellites in order to increase the position accuracy. This causes deviations from the expected system reaction towards changes of events. The introduction of GALILEO receivers, which are also able to detect GPS satellites, will solve this problem in the future. Additionally an image processing system can support the preview prediction. By means of image processing traffic lights and signs can be detected. Furthermore the status of traffic lights can be taken into account for the longitudinal control of the vehicle.

Communication between vehicles (Car2Car communication) and the infrastructure allows the system to react dynamically towards events on the driving route. As this form of communication will be handled directly between the vehicles the delay of information processing and updating, e.g. TMC, is left out. Communication devices are being specified in form of the IEEE 802.11p standard. Local hazard warning around curves and an early adaptation of the driving velocity according to the traffic ahead are applications for such communication systems.

References

[1] Donges, E.: "Ein regelungstechnisches Zwei-Ebenen-Modell des menschlichen Lenkverhaltens im Kraftfarzeug", Zeitschrift für Verkehrssicherheit 24, 1978.

[2] Pelops, www.pelops.de

[3] Zlocki, A.: "Situation classification for a traffic situation and road infrastructure adaptive ACC-system", 4th International Workshop on Intelligent Transportation (WIT) 2007, Hamburg, 20.-21.03.2007.

[4] Neunzig, D.: "Fahrerassistenzsysteme zur Verbrauchsminderung von Kraftfahrzeugen", Dissertation, Institut für Kraftfahrwesen Aachen, Aachen 2003.

[5] Honda Grant 2006: "Verkehrssituations- und infrastrukturadaptives ACC", Abschlussbericht, Institut für Kraftfahrwesen Aachen, Aachen 2007.

[6] Huang, Q.: "Satelliten-gestützte Systeme zur vorausschauenden Regelung in Straßenfahrzeugen", POSITIONs 2007, Braunschweig, 17.10.2007.

[7] Töpler, F.: "Entwicklung eines vorausschauenden ACC-Algorithmus für ein Hybridfahrzeug", Diplomarbeit am Institut für Kraftfahrwesen Aachen, Aachen 2006.

Adrian Zlocki
ika – Institut für Kraftfahrwesen Aachen
RWTH-Aachen
Steinbachstr. 7
52074 Aachen
Germany
zlocki@ika.rwth-aachen.de

Keywords: advanced driver assistance system, global navigation satellite system, sensor, digital map

Results of the EC-Project INTERSAFE

K. Ch. Fuerstenberg, B. Roessler, IBEO Automobile Sensor GmbH

Abstract

The INTERSAFE project was created to generate a European approach to increase safety at intersections. A detailed accident analysis was carried out. Based on the derived relevant scenarios driver assistant functions were developed to support the driver in critical intersection situations. These functions are described in this paper. In addition evaluation and user test results of the Intersection Driver Warning System are presented and discussed.

1 Introduction

In the 6th Framework Programme of the European Commission, the Integrated Project PReVENT includes Intersection Safety. The project started on February 1st, 2004 and successfully ended in February 2007. The INTERSAFE partners are:

- Vehicle manufacturers: BMW, VW, PSA, RENAULT
- Automotive suppliers: TRW, IBEO
- Institutes/SME´s: INRIA, ika, FCS, Signalbau Huber

The main objective of the INTERSAFE project is to improve safety and to reduce (in the long term avoid) fatal collisions at intersections. In order to identify the most relevant scenarios for accident prevention, a detailed accident analysis was carried out. Based on the scenarios and the driver mistakes derived from the accident analysis a basic functionality is described. It considers for example the time budget, which is available in order to warn the driver.

The importance of these accidents leads to a deeper analysis of the scenarios. An in depth analysis of available data from reconstructed accidents in France and Germany shows the central position of two accident types:

- Collisions with oncoming traffic while turning left and
- Collisions with crossing traffic while turning into or straight crossing an intersection.

Additionally the importance of the actual right of way regulation leads to the consideration of traffic light controlled intersections. The specification of the key technology components like sensors and communication technologies are derived from the given requirements.

Altogether, about 60%-72% of all car-related accidents in intersections are covered directly by the selection of these three scenarios. The possible coverage of other comparable accidents needs further investigations.

A general overview on the INTERSAFE subproject was given in [1]. The main focus of this paper is to give a better understanding of the functionality of the sensor based test vehicle used in this project.

2 INTERSAFE Concept & Vision

The INTERSAFE project realizes two different approaches in parallel.

The first approach is a Bottom-Up Approach, based on state of the art sensors and vehicle-to-infrastructure (V2I) communication. Furthermore, some communication modules are installed at selected intersections in public traffic to realize the bidirectional communication between the vehicle and the traffic lights. This approach results in a basic intersection system, which can be evaluated in public traffic at selected intersections.

The second approach is a Top-Down Approach, based on a BMW driving simulator. The driving simulator allows the analysis of dangerous situations, independent of any restricted capabilities of the sensors for environmental detection. The results of this approach are used to define an advanced intersection safety system, including requirements for advanced on-board sensors.

3 Demonstrator

The INTERSAFE VW Phaeton Demonstrator is equipped with two laserscanners, one video camera and additional communication systems, as shown in Fig. 1. The video camera is used to process data about lane markings at the intersection while the laserscanner collects data of natural landmarks as well as data about the road users. Highly accurate vehicle localisation is performed by fusion of the outputs of the video and laserscanner systems based on a detailed map of the intersection (high-level map).

In a second step, a dynamic risk assessment is done. This is based on object tracking and classification, communication with the traffic management and the intention of the driver. As a result of the dynamic risk assessment, potential conflicts with other road users and the traffic management can be identified.

Consequently, the intersection safety system is able to support the driver at intersections.

The laserscanner is described in detail, because this sensor detects and tracks the road users which are essential for the functionality. Nevertheless, the video processing data and the fusion with the laserscanner data is crucial for the correct functioning of the host localisation process [2].

Fig. 1. Sensor integration in the INTERSAFE demonstrator.

In the first section of this paper the localization of the host vehicle and the laserscanner based object detection, tracking and classification is described. This serves as the input level for the scenario interpretation and risk assessment which is described in the second part.

4 Localization of the Host Vehicle

GPS-based localisation is generally not able to provide sufficiently accurate or reliable localisation in urban areas where intersections are typically located. The laserscanner system obtains a relative position within the intersection by detecting landmarks such as posts and other similar fixed objects next to the intersection, which are registered in a digital map.

5 Object Detection, Tracking and Classification

In order to meet the functional requirements of monitoring crossing traffic two precise laserscanner are integrated into both left and right front corners of the demonstrator vehicle. Thus a combined scan area of 220 degree around the vehicle is achieved. Based on the laserscanner data detection, tracking and classification of road users are performed.

In the beginning the generated range profile is clustered into segments. Comparing the segment parameters of a scan with predicted parameters of known objects from the previous scan(s), established objects are recognised. Unrecognised segments are instantiated as new objects, initialised with default dynamic parameters.

In order to estimate the object state parameters a Kalman filter is well known in the literature and used in various functions, as an optimal linear estimator.

Object classification is based on object-outlines (static data) of typical road users, such as cars, trucks/buses, poles/trees, motorcycles/bicycles and pedestrians. Additionally the history of the object classification and the dynamics of the tracked object are used in order to support the classification performance.

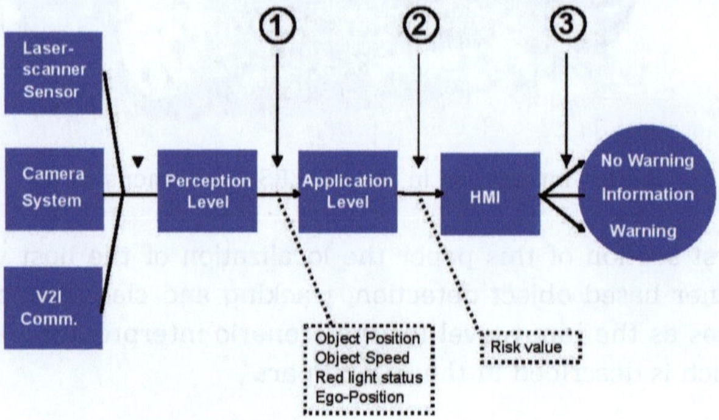

Fig. 2. General system description [3].

6 Validation Methodology

Fig. 2 shows a general diagram of an intersection safety system and the internal system interfaces connecting each system level. Examples for transferred signals are shown. The figure simplifies the architecture and depicts on which system level the tests evaluate the system performance.

Starting with the technical verification, the output of the environmental sensors (laserscanner and video) and the communication module is monitored. Here the data are compared to the values described in the technical specification. At interface 1 (perception level) the output of the relative localisation of each sensor system, the detection, tracking and classification of road users and the V2I communication is evaluated.

The second phase deals with operational verification. Here the system is inspected at interface 2 (application level). In addition, the implementation of the components is evaluated with respect to the operational specification.

Finally, the user aspects are addressed in the third part of the tests, the user test. The system is tested at interface 3 and tests with subjects are performed. The subjects performed test drives as well as the assessment of the system responses and the HMIs. Within this phase the same scenarios as in phase 2 are considered.

7 Validation Results

Some of the main validation results achieved in this project will be described in the following sections [4].

A. Sensor Test

In the first evaluation phase, the sensors were tested to check the function of the INTERSAFE system at the perception level. In order to get a representative result, diverse objects were used as sensor targets in the tests. These targets were: VW Golf (silver estate car), VW Lupo (black compact car), BMW 325i (red middle size car), BMW 728i (black large size car), Honda VFR800 (silver motorcycle), pedestrian (dark clothing) and a wooden dummy target for the test of position accuracy.

Detection Range of the Laserscanner

In this test the laserscanner maximum detection range for all five test vehicles and the pedestrian was determined. The test started with the target vehicle moving towards the standing demonstrator vehicle.

Three different approaching directions of the target vehicles were applied in this test: frontal, 45° and 90° (see Fig. 3). If the opponent vehicle is perpendicular and far away from the host vehicle, it will typically leave the intersection before the host vehicle enters. Therefore the maximum detection range of perpendicular vehicles was just additional and performed with the VW Golf.

In order to avoid coincidences with regard to repeatability, every test was carried out twice for each target and approaching direction.

The maximum detection ranges of all the frontal and 45° tests are illustrated in Fig. 4. The CAN specification in the VW Phaeton demonstrator was limited to 200 m; a higher distance was not necessary.. Therefore in this test, the maximum detection range of the laserscanner reported is 200 m.

During the tests, all the vehicles were correctly classified immediately after detection. All the cars were detected at a distance of more than 200 m, both in frontal and 45° tests. Just the maximum detection range of the motorcycle was slightly shorter.

The detection of pedestrian was also tested. Result showed that the pedestrian was detected and correctly classified as a pedestrian at distances of 110 m. In the 90° tests, the VW Golf was detected at about 165 m.

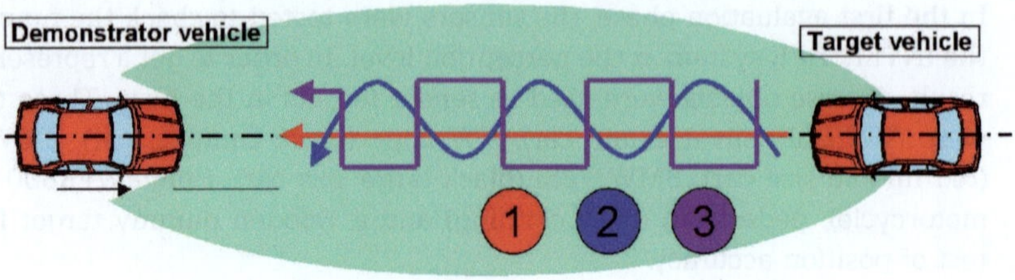

Fig. 3. Test layout.

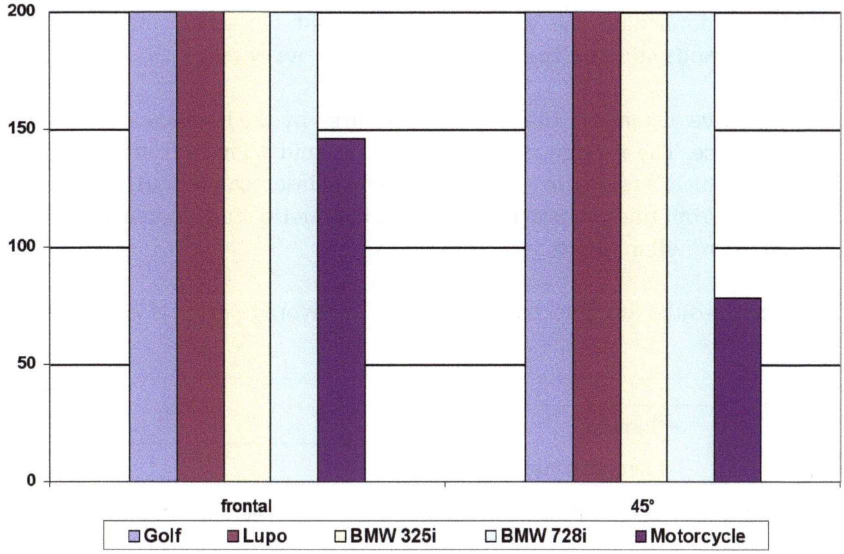

Fig. 4: Detection range of the Laserscanner.

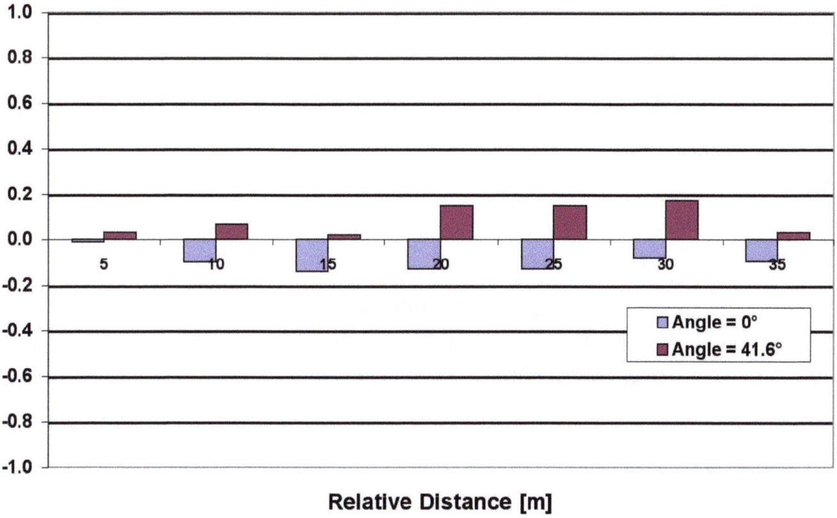

Fig. 5. Object range accuracy of the Laserscanner.

Object Range Accuracy of the Laserscanner

A wooden dummy is used as the sensor target in this test. It is positioned either directly in front of the demonstrator vehicle or in a 45° position. In order to determine the accuracy at different distances, the demonstrator vehicle stands still and the target is moved in every test.

The relative distance of the target is measured by the laserscanner and a reference device. The reference line is located around 15 m in front of the demonstrator vehicle's rear axle (the zero axis of the laserscanner output). The target is moved from this position until 50 m in five metre steps; namely 20 m, 25 m, 30 m, 35 m, 40 m, 45 m and 50 m.

The test results are illustrated in Fig. 5. The average error is around 10 cm.

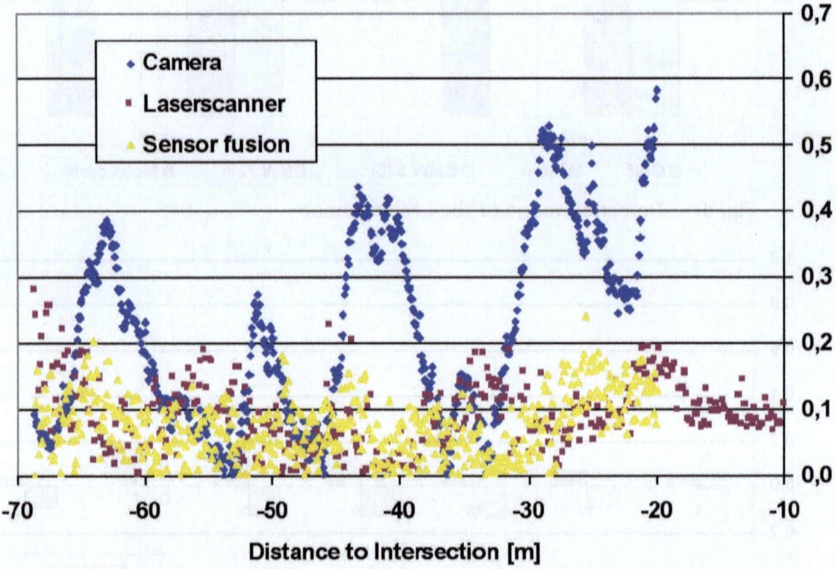

Fig. 6. Localisation error over distance for the camera (blue), laserscanner (red) and the fusion system (yellow).

Localisation Accuracy

This test was applied to inspect the Localisation accuracy of the laserscanner, the video system and the fusion output. The laserscanner system localises the vehicle position by detecting landmarks while the video system utilises road markings. As an example, the test for evaluation of longitudinal distance was described here.

In order to determine the demonstrator vehicle's distance to the intersection, a microwave sensor and a light barrier sensor were mounted at the rear end of the car. Four reflectors were put on the ground as reference positions for the light barrier. The distances to the intersection of these references were 10 m, 30 m, 50 m and 70 m respectively. According to the vehicle speed measured by the microwave sensor and the reference positions by the light barrier, the vehicle position could be calculated precisely.

The test was carried out five times. The localisation error for one exemplary test run is shown in Fig. 6. The average absolute errors of each test as well as the average of all the tests are summarised in Fig. 7. The outputs of all localisation systems were continuous, meaning no signal drop-outs occurred.

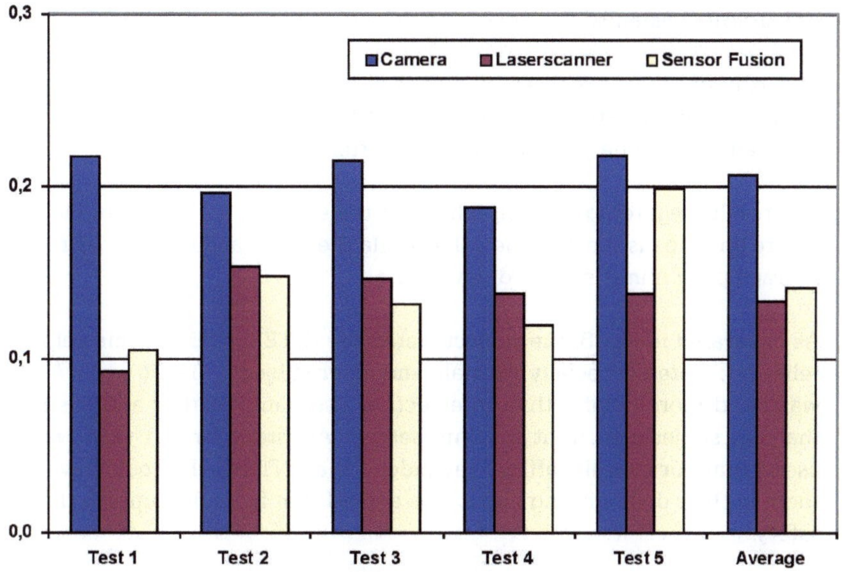

Fig. 7. Localisation accuracy distance for the camera (blue), laserscanner (red) and the fusion system (yellow).

B. System Test

The INTERSAFE system functions at the application level were tested in this evaluation phase. The evaluations were carried out based on the number and the rate of correct alarms, false alarms and missing alarms.

During the test, the demonstrator vehicle was driven through the intersection and the functions of the system were evaluated in the form of a check list.

The results of system test indicated that the Intersection Assistant had a correct alarm rate of 93% in left turn scenarios and 100% in lateral traffic scenarios. Both traffic light assistant systems achieved together an average correct alarm rate of 90%.

C. User Test

Sixteen subjects had been selected by taking their age, gender and driver experience into account. Each subject took around 2.5 hours to drive the demonstrator vehicles on ika's test track and assessed the performance of the INTERSAFE systems.

The assessment was realised by means of questionnaires. Subjects were asked to fill out a pre-questionnaire before the driving test, three questionnaires during the test and one post-questionnaire after the test.

The results regarding the helpfulness of the systems are shown in Fig. 8. The centre line means the average value while the lower and upper lines mean the average plus/ minus standard deviation.

As illustrated in Fig. 8, the subjects rated the INTERSAFE systems helpful and relieving, stated especially by male and older subjects. Traffic Light Assistant was rated more helpful than Intersection Assistant. Further analysis showed that the subjects thought the Intersection Assistant for left turn was more useful than for lateral traffic. They judged that INTERSAFE could have helped them in their daily driving and it was agreed that it would improve the traffic safety.

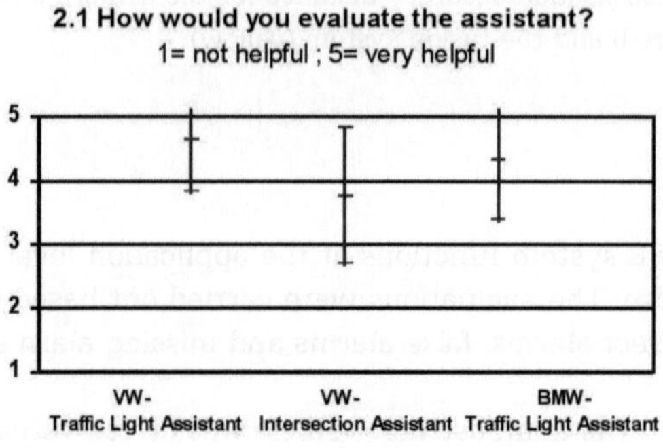

Fig. 8. Helpfulness of the Intersection Assistant systems.

8 Conclusions

This paper describes the evaluation and testing of the two INTERSAFE demonstrator vehicles developed and implemented for intersection driver assistance. The tests were carried out in three evaluation phases: sensor test, system test and user test.

During the sensor test, sensors (laserscanner and video system) and the communication module were under investigation. The focus of this phase was to verify the sensor performance required for the INTERSAFE functions. The chosen sensors were fully suitable to fulfill the tasks in the INTERSAFE Intersection Driver Assistance System.

The system test was carried out to check the system's functionality before the user test. Because of the system layout, a false-alarm-free result had not been achieved. In average, the correct alarm rate of the intersection assistant system was 97% and for the traffic light assistant 90% for both demonstrators.

In the last testing phase, sixteen subjects had driven the demonstrator vehicles on the test track and assessed the performance of the INTERSAFE Intersection Driver Assistance Systems. Generally (and in particular, male and older subjects) think the INTERSAFE systems are helpful and relieving. Subjects think Traffic Light Assistant is more helpful than Intersection Assistant. Intersection Assistant for left turn scenarios is rated more useful than the one for lateral traffic scenarios.

The INTERSAFE systems would have helped the subjects in their daily driving and it was agreed that it would improve traffic safety and that they want to have them in their car.

References

[1] Fuerstenberg, K. Ch.: Intersection Safety - The EC-Project INTERSAFE. Proceedings of ITS 2005, 12th World Congress on Intelligent Transport Systems, ITS 2005 San Francisco, USA.

[2] Heenan, A.; Shooter, C.; Tucker, M.; Fuerstenberg, K. Ch.; Kluge, T.: Feature-Level Map Building and Object Recognition for Intersection Safety Applications. Proceedings of AMAA 2005, 9th International Conference on Advanced Microsystems for Automotive Applications, March 2005, Berlin, Germany.

[3] "INTERSAFE Validation plan", INTERSAFE Deliverable D40.77, Brussels, 2006.

[4] "INTERSAFE Evaluation results", INTERSAFE Deliverable D40.71/2/3/4, Brussels, 2007.

Bernd Roessler
Research Department
IBEO Automobile Sensor GmbH
Merkurring 20
22143 Hamburg
Germany
bernd.roessler@ibeo-as.com.

Kay Ch. Fuerstenberg
Director of Research
IBEO Automobile Sensor GmbH
Merkurring 20
22143 Hamburg
Germany
kf@ibeo-as.com.

Keywords: intersection safety, advanced driver assistance system, vehicle demonstrator, laserscanner, sensor tests

Powertrain

Crank Angle Resolved Determination of Fuel-Concentration and Air/Fuel Ratio in a SI-Production Engine by Using a Modified Optical Spark Plug

A. Grosch, V. Beushausen, Laserlaboratorium Göttingen e.V.
O. Thiele, LaVision GmbH

Abstract

Local, crank angle resolved fuel density variations were detected with a fiber optical sensor system in the vicinity of the spark position in a cylinder of a four-stroke SI production engine. The infrared absorption sensor detects the attenuation of infrared radiation in the 3.4 µm wavelength region due to the infrared vibrational-rotational absorption band of hydrocarbons (HC). The absorption path was integrated in a modified spark plug and a tungsten halide lamp was used as an infrared light source. All investigations were carried out on a four-stroke spark ignition engine with fuel injection into the intake manifold. The engine operated with common gasoline (Euro Super) at different preset overall air/fuel-ratios. Fuel concentration variations and the mixing process before the combustion were investigated for single engine cycles. In order to derive air/fuel-ratios from the measured HC-concentrations a special approximation method was used.

1 Introduction

Modern ignition engine development targets at an enhancement of engine performance and a simultaneous reduction of pollutant emissions. A continual improvement of the combustion process in engines is necessary in order to meet these oppositional requirements. The ignition and an efficient combustion of the air/fuel-mixture are mainly influenced by the fluctuating air/fuel-ratio during the compression stroke at the position of the spark plug. The mixture formation is dominated by strongly interacting processes and flow parameters like fuel-injection, vaporization, in-cylinder flows, pressure and temperature. These factors of influence can lead to strongly fluctuating fuel density at the ignition position which has strong influence on combustion performance. A highly time-resolved measurement method for analysing the fuel density at the spark position is of great importance to optimize combustion performance. In order to optimize the combustion performance, engineers need to know the fuel density and the air/fuel-ratio especially at the ignition location and if possible crank angle resolved. Additionally this information needs to be generated in production engines with negligible mechanical modifications and with common gasoline as fuel because changes in engine geometry and fuel composition inevitably lead to alterations in the mixture formation and the following combustion processes.

Recently different optical methods were applied in order to analyze mixture formation and combustion processes inside the cylinder of car engines. Especially diagnostic methods like Rayleigh scattering [1, 2], Raman scattering [3]-[5], and Laser Induced Fluorescence LIF [6-9] have been used. These methods usually need to be carried out at so called transparent research engines which are strongly modified for optical access. However large optical access reduces the mechanical stability of the cylinder and its thermodynamical properties. Besides, these optical techniques often demand the use of special model fuels which usually can not reproduce all properties of common gasoline.

Another method, applied for mixture formation and combustion analysis, is gas sampling [10-12]. Gas sampling is the best known method for crank-angle-resolved in-cylinder fuel concentration measurements in production engines despite the time delay caused by the length of the sampling pipe.

In several investigations infrared absorption is used for measuring the hydrocarbon concentrations in the cylinder. Hall et al. [13-18] used a tungsten halide lamp as an infrared light source with a corresponding band-pass filter that coincides with the fundamental vibrational-rotational absorption band of hydrocarbons at 3.4 µm. A HeNe-Laser was used by Tomita et al. [19-22]. This laser emits radiation at a wavelength of 3.392 µm which is also suitable for fuel concentration measurements. Usually a calibration with several air/fuel-mixtures at different temperatures and pressures is needed to eliminate the temperature and pressure dependence of the monochromatic absorption coefficient.

Unlike to most of the other measurement techniques the infrared absorption method allows a crank angle resolved determination of the fuel concentration in many consecutive cycles and the sensor can be integrated in a modified spark plug, so that the combustion chamber has not to be modified for optical access. An outstanding advantage of this technique is that common gasoline can be used as the fuel, since no interferences from fluorescence emissions need to be worried about.

In this study, a tungsten halide lamp was used as an infrared radiation source. All investigations were carried out on a production four stroke engine, operating with standard gasoline (Euro Super) and fuel-injection into the intake manifold. The engine worked with different preset overall air/fuel-ratios (lambda values) which occur at the end of the compression stroke in the combustion chamber due to the homogeneous engine running mode. All investigations were carried out at low engine revolution speed (500 rpm) in order to simulate starting conditions.

Fuel density variations in motored and fired single engine cycles were investigated. The fuel density and pressure data resulting from optimal combustion as well as particularities like delayed combustion and misfire are displayed in this study.

Furthermore, time resolved mixture-formation near the ignition position was investigated in this study. The mixture generation with different quantities of fuel and its effects on combustion was detected crank angle resolved during compression. An approximation method was applied in order to calculate lambda-values (air/fuel-ratios).

2 Infrared Absorption Method

Due to the fundamental CH stretch vibrational band, IR-radiation is strongly absorbed by fuel hydrocarbon compounds at wavelength around 3.3 and 3.5 µm. Hardly any other combustion related molecular species has absorption bands in this wavelength region which interfere with those of hydrocarbons. Only water molecules absorb radiation in the named wavelength range but due to low molecular density and lower absorption strengths the absorption is about hundred times weaker than the absorption caused by hydrocarbons [23].

The concentration of hydrocarbons determines the intensity I of monochromatic radiation which passes the measurement region. There is a correlation between transmissivity I/I_0 and molar concentration C of fuel which is expressed by Lambert-Beer´s law (Eq. 1). I_0 describes the intensity for the case that no hydrocarbons are present inside the absorption path. During a single engine cycle the intensity I_0 can be measured at several times. This is important since I_0 should be recorded most often, best once every cycle, in order to detect and correct fouling of the in-cylinder optics. For every engine, a time interval exists where no fuel is present in the absorption path at the spark plug. Usually this is the case during the exhaust phase when all fuel molecules should be burned. However in case of misfires as well as delayed or incomplete combustion, unburned fuel is still existent during the exhaust phase, and it is not possible to detect the ongoing I_0 in the exhaust gases. For that case the intensity I_0 can be measured during the intake stroke, after the air/fuel mixture from the intake manifold has passed the measuring area and only pure air is flowing through the absorption path. In Lambert-Beer´s law L defines the length of the absorption path.

$$\frac{-\ln\left(\frac{I}{I_0}\right)}{\varepsilon_\lambda \cdot L} = C \tag{1}$$

The absorption of monochromatic radiation usually depends on temperature and pressure of the absorbing gas. Therefore the wavelength dependent absorption coefficient ε_λ of the fuel molecules is not constant during a complete engine cycle due to Doppler broadening of the absorption lines during compression. In the case of gasoline another effect arises which superposes Doppler broadening. Common gasoline is composed of a huge amount of different hydrocarbon species whose CH-fundamental stretch vibrations exhibit slight energy shifts due to different molecular structures and the resulting shifts of the vibrational energy levels. At large these shifts result in a broad absorption band of gasoline (3.3-3.5 µm) which is a superposition of all the absorption bands of the single fuel components. Since this broadening effect is dominating, Doppler broadening does not carry weight any more.

A broad band radiation source and an optical band pass filter which matches the whole absorption band of the fuel-hydrocarbons were used for the experiments. Hall et al. verified that Doppler broadening of the infrared absorption lines, caused by temperature and pressure variations during compression in an engine cycle, can be neglected. Therefore the transmission curve of the band pass filter has to match the whole absorption band including the broadening effects. The absorption coefficient depends on the wavelength across the absorption band and the resulting band integrated absorption coefficient is not following Lambert-Beer´s law directly [16].

In case of a closed volume during the compression phase, the number of absorbing fuel molecules as well as the number of air molecules is constant. This means there is a constant concentration of fuel in the volume of air. A decrease of the cylinder volume increases merely the number of absorbing molecules in the absorption path, which decreases the transmissivity I/I_0. Therefore the transmissivity is dependent on the density of fuel molecules in the alternating cylinder volume. Since there are no corresponding data for band integrated fuel absorption coefficients available in the literature and the changing fuel density has to be related to a corresponding transmissivity signal a calibration has to be carried out.

3 Experimental Setup

A tungsten halide lamp (75 W) was used as a broadband radiation source. An elliptical reflector focuses the radiation onto the end of a sapphire fiber with a diameter of 425 µm. Sapphire fibers are used because they are transparent in the mid-infrared spectral region and do simultaneously withstand high temperatures which occur after ignition at the spark plug.

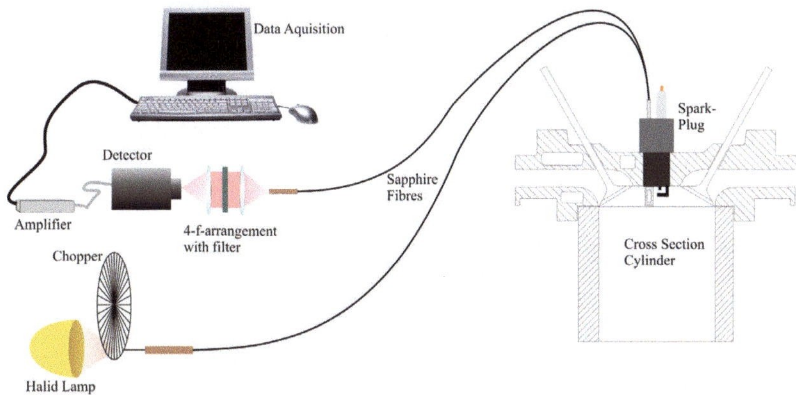

Fig. 1. Experimental setup

A square wave signal is produced by a chopper wheel, positioned between lamp and fiber. The square wave signal is necessary in order to correct errors caused by infrared background radiation generated by the excited hydrocarbon molecules during compression and the radiation of the heated metal mirror. The rotational frequency of the chopper wheel has to be high enough to generate a crank-angle resolved data flow. For an engine revolution speed of 500 rpm a square wave frequency of around 3 kHz is necessary to calculate absorption signals with a temporal resolution of one degree crank angle.

The sapphire fiber transmits radiation to the absorption path which is placed at the tip of the modified spark plug (see Fig. 2 and Fig. 3) and a small spherical mirror reflects the radiation back to a second fiber which guides the light to the detector. The absorption path consists of a small stainless steel tube with integrated slits in order to enable free circulation of fuel/air mixture (Fig. 3).

The measured signal comprises two kinds of different information. If there are hydrocarbons in the absorption path the intensity of the square wave signal will decrease. This reduced signal is an indication of the fuel density in the measurement region. Additionally background radiation at 3.4 µm is detected, which results from thermal radiation emitted by hot cylinder walls, hot mirror surfaces and the spontaneous emission of compression heated fuel. This radiation superimposes the absorption signal. By subtracting the pure background signal, which is detected during the chopper phase "excitation light off" from the signal detected at the chopper phase "excitation light on", it is possible to calculate the pure hydrocarbon absorption signal.

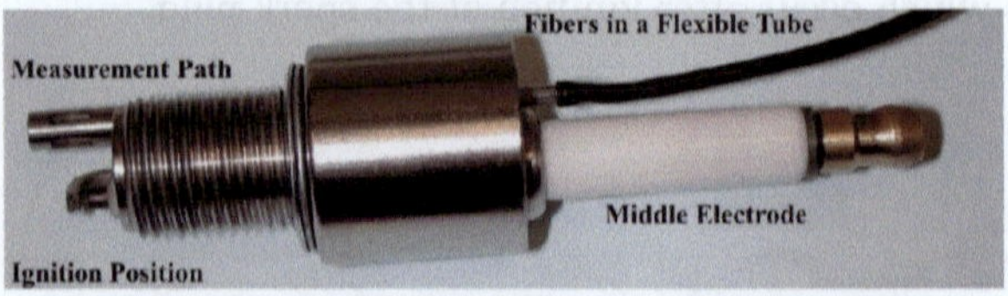

Fig. 2. General view of the spark plug with integrated absorption path

Fig. 3. Spark plug head with eccentrically placed ignition electrodes and absorption path

The ignition electrode of the modified spark plug is positioned eccentrically beside the absorption path as shown in Fig. 3, in order to avoid electrical flashovers. More technical details about the construction of the measurement spark plug are presented in a previous publication [24]

The second fiber transmits the radiation from the measurement region back to a HgCdTe-detector. A band-pass filter with a center wavelength of 3.42 µm and a FWHM of 300 nm blocks the broadband radiation of the tungsten halide lamp and the background radiation from hot mirror. Only the radiation in the spectral range of the C-H absorption band can pass the filter.

A voltage amplifier with an integrated low pass filter, placed between the detector and the data acquisition system, reduces the signal noise. Beside the absorption signal, the in-cylinder pressure was recorded by a Kistler pressure conductor with sub crank angle resolution

The engine, used for the experiments, was a throttled four stroke spark ignited production engine with fuel injection into the intake manifold. Tab. 1 shows the engine specifications and the opening and closing times of the intake and the exhaust valves.

Parameter	*Value*
Stroke / mm	*85*
Bore / mm	*89*
Compression Ratio	*9:1*
Intake Valve Opening (IVO) ***Intake Valve Closing (IVC)***	*20° after TDC of intake* *249° after TDC of intake*
Exhaust Valve Opening (EVO) ***Exhaust Valve Closing (EVC)***	*510° after TDC of intake* *20° after TDC of intake*
Intake Pressure	*0,520 bar*
Intake Temperature	*303 K*
Coolant Temperature	*353 K*

Tab. 1. Engine Specifications

3 Measurements

3.1 Raw Data

The raw data recorded during a fired engine cycle consist of a square wave signal with a frequency of 3 kHz which is shown in Fig. 4 and 5. The amplitude of the upper signal envelope is a superposition of transmitted signal from the tungsten halide lamp and background emission out of the cylinder. The lower signal envelope represents pure background radiation mainly due to hot surfaces and spontaneous emission of heated hydrocarbons. The graph displays a complete cycle of 720 degree crank angle, beginning at Top Dead Center of intake (TDC).

At roughly 20 degree crank angle, the pressure curve decreases due to closure of the exhaust valves and the volume expanding movement of the piston. The throttle limits the air flow from the intake manifold and generates a pressure of 520 mbar inside the cylinder. At point of injection an air/fuel-mixture cloud flows into the cylinder, which is noticeable as a small decrease of the IR-signal around 40 to 90 degree crank angle. The square wave signal is diminished due to the absorption caused by the excitation of CH-stretch vibrations of the gaseous fuel, which flows into the cylinder and through the absorption path at this time. The background radiation of the hot mirror decreases because the IR-radiation is shadowed by cool fuel droplets and absorption by cool vaporized fuel.

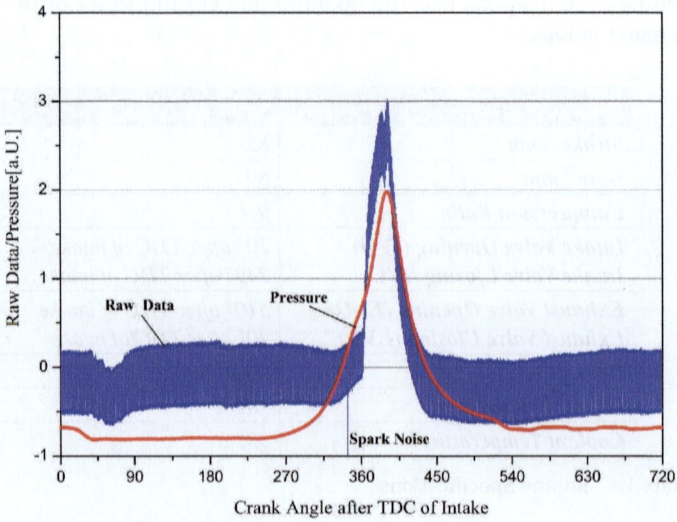

Fig. 4. Raw absorption data and pressure trace for one single engine cycle

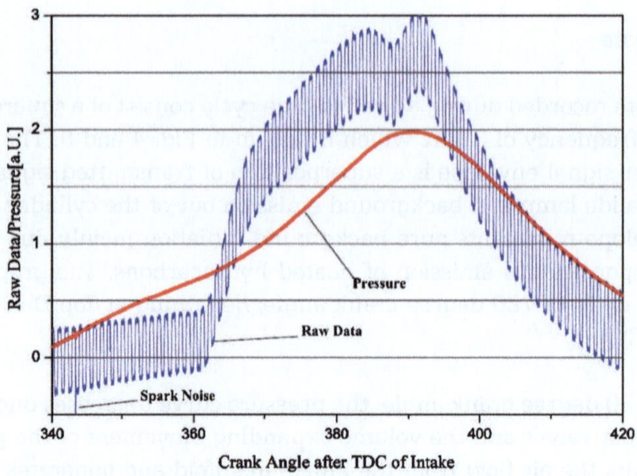

Fig. 5. Enlarged view of the IR-transmission raw data

During compression the pressure increases and the amplitude of the square wave signal (difference between "excitation light on" and "excitation light off") decreases due to increasing density of absorbing hydrocarbons. The spike in the transmission signal at roughly 343 degree crank angle indicates the spark noise of the ignition.

Fig. 5 shows an enlarged view of the raw data of Figure 4. This data representation impressively demonstrates the high data acquisition frequency with one degree crank angle resolution which can be generated by the experimental setup. At least one full signal period (signal on and signal off) during one degree crank angle is needed in order to calculate the crank angle resolved transmission signal.

An immense increase of the background radiation during the power stroke occurs at crank angles around 362°. This is due to strong increase of IR-radiation from combustion the flame as well as increased thermal radiation of the metal mirror inside the absorption path which is also heated by combusting gases and eventually from thermal radiation of glowing soot particles. It is obvious in Fig. 5 that the increase in background radiation occurs at a time delay of about 19 degrees crank angle relative to ignition at about 343 degrees crank angle. This delay results from the time needed of the developing flame kernel to bridge the gap between ignition electrode and absorption path.

3.2 Calibration Procedure

In order to convert the measured absorption intensities to quantitative fuel densities during the whole combustion cycle a calibration with premixed air/fuel-mixtures of well known concentration need to be carried out. Unfortunately the used engine test bench does not offer the possibility to generate a defined, homogenous, vaporized gasoline/air-mixture. However, since the fuel injection occurs at an early stage of the engine cycle into the intake manifold the generated air/fuel-mixture has time to homogenize during intake and compression so that it can be assumed to be homogenous at the end of the compression stroke. In case of misfire, a homogeneous mixture is established over a longer period of time up to the point where the intake valves open again. This leads to another possibility to calibrate the absorption signals of the sensor on fuel densities by using the homogenous mixture after a misfired cycle. During a misfired cycle, an adiabatic compression and expansion of the in-cylinder gas mixture can be assumed when the valves are closed. For a first approximation temperature losses will be neglected and the equations for ideal gases will be used to describe the gas-mixture during compression.

The in-cylinder pressure p_{cyl} was recorded by a piezo pressure conductor. It is possible to calculate the adiabatic temperature variation during a motored engine cycle by using the recorded pressure p_{cyl} in Eq. 2. The temperature T_{in} and pressure p_{in} are the values measured in the intake manifold. For the present mixture the specific heat ratio κ equals 1.3.

$$T_{cyl} = \left(\frac{p_{cyl}}{p_{in}} \right)^{\frac{\kappa-1}{\kappa}} \cdot T_{in} \tag{2}$$

The amount of substance of the air/fuel-mixture $n_{a/f}$ can be calculated by the ideal gas law (Eq. 3) at the time the intake vales are closing ($p_{cyl\text{-}IVC}$/$V_{cyl\text{-}IVC}$/T_{cyl}) and this calculated number $n_{a/f}$ will be constant during the compression.

$$n_{a/f} = \frac{p_{cyl-IVC} \cdot V_{cyl-IVC}}{R \cdot T_{cyl-IVC}} \tag{3}$$

The λ-value (Eq. 4) of the homogenous mixture depends on the injected quantity of fuel. This value can be controlled by the lambda probe which measures the concentration of oxygen remaining in the exhaust gas for multiple engine cycles. It describes the mass ratio of the actual air/fuel mixture compared to the stoichiometric one which indicates complete combustion. The mass ratio of the air/fuel-mixture for stoichiometric combustion is assumed to be 14.7:1. The masses in Eq. 4 can therefore be replaced by a product of the amount of substances and the molar masses.

$$\lambda = \frac{m_{air} / m_{fuel}}{\left(m_{air} / m_{fuel} \right)_{sto}} = \frac{n_{air} \cdot M_{air}}{14.7 \cdot n_{fuel} \cdot M_{fuel}} \tag{4}$$

Eq. 4 can be solved for the number of air molecules n_{air} (Eq. 5). In case of the detected misfired cycle the engine works with a homogeneous mixture of a specific lambda value of $\lambda = 1.1$. The amounts of substances of air and fuel can be calculated by following equations 5 and 6. Whereas the molar masses of air M_{air} and fuel M_{fuel} are 28.97 g/mol and 100 g/mol respectively.

$$\lambda \cdot 14.7 \cdot n_{fuel} \cdot \frac{M_{fuel}}{M_{air}} = n_{air} \tag{5}$$

$$n_{a/f} = n_{fuel} + n_{air} \tag{6}$$

As mentioned before, the gas mixture is assumed to be homogenous during the expansion after TDC. In case of equilibrium, the total pressure of a gas mixture is the sum of the partial pressures of each individual gas in the mixture (Eq. 7).

$$p_{a/f} = p_{air} + p_{fuel} \tag{7}$$

In an air/fuel mixture, which is assumed to be an ideal gas, each component has a partial pressure which the gas component would have in case it would occupy the cylinder-volume alone. The amounts of substances of the mixture $n_{a/f}$ and of the fuel n_{fuel} are known and constant during compression and expansion. Eq. 8 determines the partial pressure of fuel p_{fuel} in the cylinder during the compression with the measured in cylinder pressure p_{cyl}.

$$p_{fuel} = \frac{n_{fuel}}{n_{a/f}} \cdot p_{cyl} \tag{8}$$

The knowledge of the time resolved partial pressure of fuel p_{fuel} and air p_{air}, the referring adiabatic temperature T_{cyl} and the gas constant R, allows to calculate the crank angle resolved density of the homogenously distributed fuel, ϱ_{fuel} and air ϱ_{air}, in the cylinder during compression (Eq. 9 and 10).

$$\rho_{fuel} = \frac{p_{fuel} \cdot M_{fuel}}{R \cdot T_{cyl}} \tag{9}$$

$$\rho_{air} = \frac{p_{air} \cdot M_{air}}{R \cdot T_{cyl}} \tag{10}$$

The calculated fuel densities which occur during expansion can be plotted over the measured transmissivity I/I_0 as shown in Fig. 6. The data were fitted with a rational function of order two, which is indicated by the solid line. By using this function it is possible to calculate any fuel density, referring to the measured transmissivity.

As mentioned before, this calibration procedure was performed with data generated during a misfired cycle with a lean air/fuel mixture of $\lambda = 1.1$ during the expansion stroke. At richer mixtures the density of fuel is higher at the end of the compression stroke. The data for measurements performed with rich mixtures at $\lambda = 0.7$ exhibit a transmissivity I/I_0 of about 0.65. It is important for the interpretation of the following graphs to keep in mind that the calibration only covers a transmissivity range of 0,72 to 0,98. All data beyond this range are calculated values gained by the fitted curve.

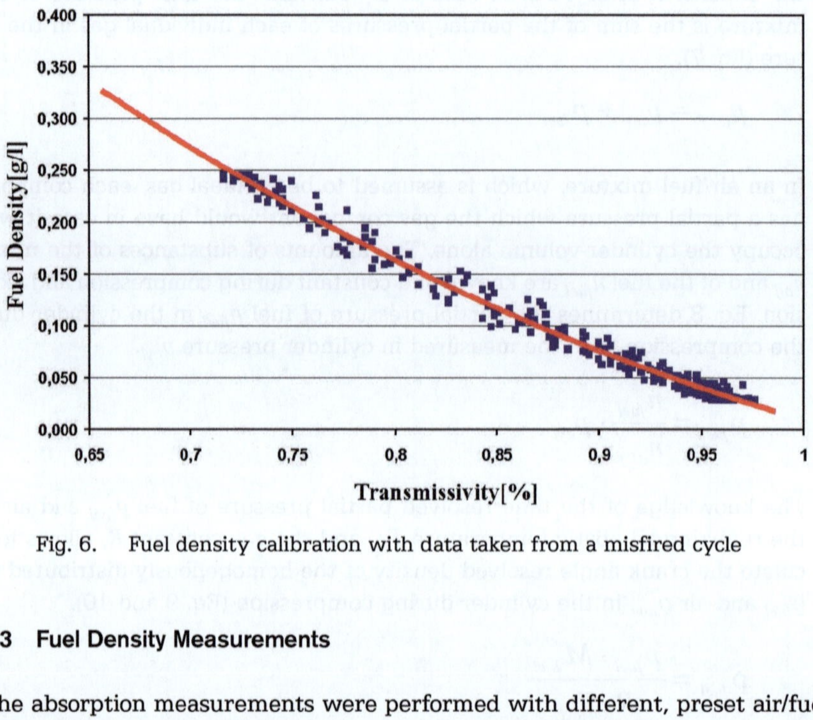

Fig. 6. Fuel density calibration with data taken from a misfired cycle

3.3 Fuel Density Measurements

The absorption measurements were performed with different, preset air/fuel-ratios in the range between $\lambda = 0.7$ and $\lambda = 1.1$. One single data set for each air/fuel-ratio contains 50 single cycles. Representatively, the measurements made with an air/fuel-ratio of $\lambda = 1.1$ will be discussed in more detail because the data fit best in the calibrated range (see chapter 3.2) and the extreme ratios seem to be interesting because the engine operation shows instabilities at these air/fuel-ratios. Fig. 7 shows the measured fuel densities for 50 single engine cycles and a preset λ-value of 1.1. The first peak between 40 and 90 degree crank angle indicates fuel, flowing from the intake manifold into the cylinder and crossing the absorption path. It is visible, that cyclic fluctuations of the fuel density during intake are quite strong. This may be due to incomplete evaporation inside the intake manifold. Fuel droplets could cause errors in the absorption measurements, therefore only qualitative conclusions can be made for the intake phase.

After the fuel has passed the measurement region the detected fuel density decreases rapidly and only pure air flows into the cylinder. Between 100 and 180 degree crank angle hardly any fuel is detectable, only noise, resulting from the detector is visible.

The compression starts at 180 degree crank angle and the piston pushes the air/fuel-mixture back to the cylinder head to the measurement spark plug, which leads to an increase of the measured fuel density. After combustion, between 360 and 540 degree crank angle, strong signal intensity fluctuations occur. The combustions sometimes seem to be incomplete at this time. The variation of fuel densities during this period of the engine cycle will be discussed for special cycles later on. Pre-combustion fuel density fluctuations during the compression phase will be discussed first.

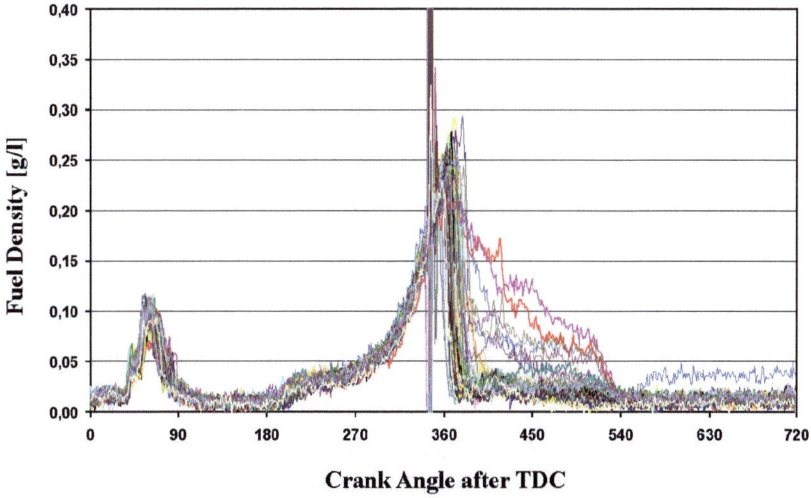

Fig. 7. 50 crank angle resolved fuel density curves for a preset air/fuel-ratio of $\lambda = 1.1$

Fig. 8 displays the standard deviations of the recorded data sets at air/fuel-ratios of $\lambda=1.1$ / $\lambda=1.0$ / $\lambda=0.7$ respectively. The fluctuations of the fuel densities during cycles, recorded at rich or lean engine operation modes ($\lambda=1.1$ and $\lambda=0.7$), are directly compared to data, recorded for a preset stochiometric air/fuel-mixture with a lambda value of $\lambda=1.0$.

During intake, at roughly 60 degree crank angle, the fuel density signals vary stronger in case of rich mixtures with $\lambda=0.7$. This stronger variation is a result of the fuel surplus because especially rich air/fuel-mixtures with a large amount of fuel lead to an incomplete vaporization at this time in the cycle. The fluctuations caused by fuel droplets crossing the absorption path are assumed to be higher at rich mixtures, compared to lean mixtures.

Between 90 and 180 degree crank angle, after the fuel from the intake manifold has passed the measurement region, there are hardly any fluctuations for all investigated ratios. This is also an indication for the assumption that no fuel molecules are present and only pure air flows through the absorption path at this time. During compression, the standard deviation for every investigated ratio increases and indicates more fuel density fluctuations. Especially at the end of compression, between 270 and 340 degree crank angle, the fuel density fluctuations of leaner and richer mixtures are higher compared to the stochiometric one.

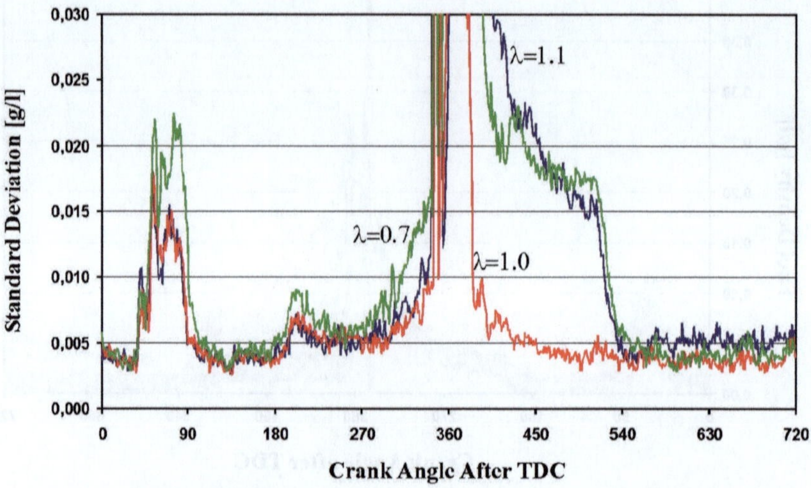

Fig. 8. Crank angle resolved standard deviation of the fuel density for different air/fuel-ratios

The raw data of the single cycle fuel densities show strong cyclic variations for crank angles around 360 degree crank angle. This is due to strong fluctuations in the flame kernel development which leads to fluctuations in the time, the flame front needs to reach the absorption path. This phenomenon occurs independently for any preset air/fuel mixture. Nevertheless it tends to be stronger for mixture ratios which vary more from the stoichiometric one. The strong variations up to 380 degree crank angle are mainly due to strong combustion fluctuations.

The fuel density variations, caused by incomplete or delayed combustion at rich or lean mixtures, are significant in Fig. 8 between 380 and 540 degree crank angle. After 540 degree crank angle there is hardly any fuel density fluctuation detectable. Slight differences in the fuel density variation during the outlet stroke result from unburned fuel which is pushed through the absorption path into the outlet manifold by the movement of the piston.

Fig. 9 shows the fuel density over crank angle for 50 averaged single engine cycles at different preset air/fuel-ratios. As described before, the first peak indicates the fuel flowing through the absorption path during intake. The amplitude of the signal is directly related to the amount of fuel, which was injected into the intake manifold. Richer mixtures lead to a lower transmissivity and therefore to a higher density of absorbing fuel molecules.

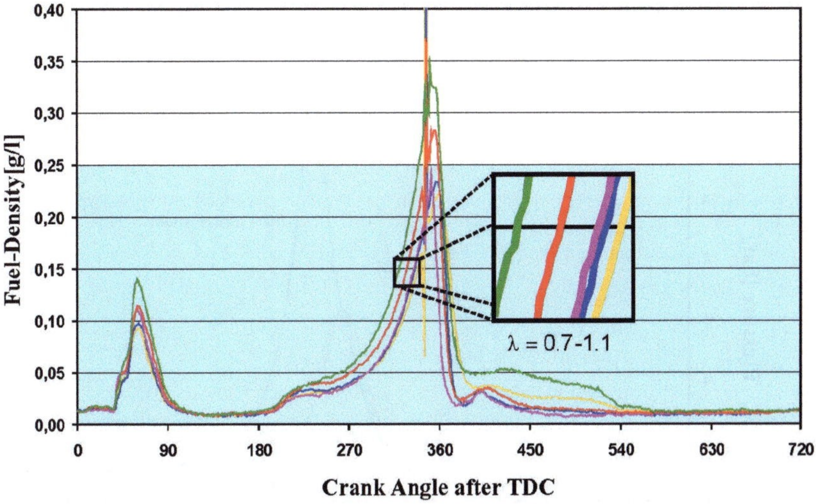

Fig. 9. Fuel density measurements at different air/fuel-ratios

During compression the curves slightly differ from each other, depending to the preset air/fuel-mixture. The enlarged region before the ignition shows the differences between the fuel densities according to the preset lambda values. The more left a curve lays the richer is the mixture. The highlighted region in the graph indicates the calibrated density range. During compression, most of the fuel should be evaporated, hence the density values can be analyzed quantitatively.

Engine cycles with an ideal and a delayed combustion as well as a misfired cycle are displayed in Fig. 10 and 11. The graphs show related pressure and fuel density data of single engine cycles at a preset lambda value of $\lambda = 1.1$.

All pressure curves in Fig. 10 show similar behavior during the intake and compression stroke. At roughly 20 degree crank angle the exhaust valves close and the movement of the piston induces a pressure drop. At 249 degree crank angle the intake valves close, the compression of the gas mixture starts and therefore the pressure rises. The curves obviously differ strongly after TDC of compression. The pressure of the misfired cycle decreases during the power stroke nearly symmetrically to the compression stroke. During the fired cycle, the highest pressure occurs some degrees crank angle after TDC. The pressure curve of a delayed combustion is characterized by a second increase of the pressure after TDC. There are some pressure fluctuations detectable during the exhaust stroke due to pressure variations in the exhaust pipe.

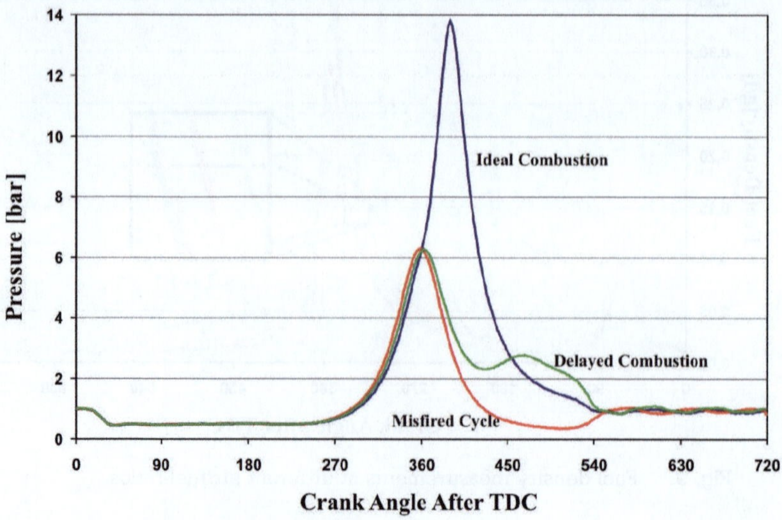

Fig. 10. Pressure curves for single engine cycles

Fig. 11 displays the fuel density belonging to the pressure curves in Fig. 10. All fuel concentrations show very similar behavior during the intake and compression strokes. The hydrocarbon-density increases rapidly after the intake valves open and the vaporized fuel flows from the intake manifold into the cylinder. The fuel density decreases back to the initial value after the fuel has passed the absorption path of the sensor because only air is entering the cylinder from the intake manifold. All fuel densities rise in the same way during the homogenization phase of the compression stroke.

The fuel density characteristic of the misfired cycle indicates a homogenous mixture during the power stroke because the density decreases symmetrically to the increase during the compression stroke. At 510 degree crank angle the exhaust valves open and exhaust gas from the preceding cycle flows back into the cylinder, due to the resulting pressure difference between the cylinder volume and the outlet manifold. This leads to a further decrease in measured fuel density. At 540 degree crank angle, the density rises again because the piston pushes the unburned mixture, which is not yet mixed with the recirculating exhaust gas, out of the cylinder. The constant value of the fuel density during the whole exhaust stroke is an indication of a homogenous mixture as well.

During the well ignited cycle, fuel density drops down after the flame front has passed the measurement volume because there are no detectable hydrocarbons in the measurement region anymore. The small increase of fuel density at 420 degree crank angle indicates absorption probably caused by hydrocarbons outgassing from crevices inside the cylinder [16].

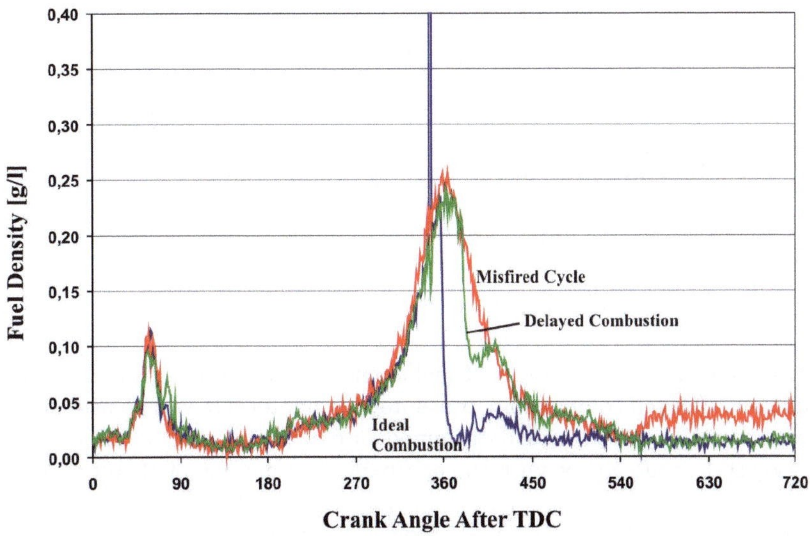

Fig. 11. Fuel density measurements for single engine cycles

The ignition of the delayed cycle occurs also at roughly 343 degree crank angle indicated by the spark noise. Fig. 11 also shows the curve of the fuel density during the delayed combustion. At 380 degree crank angle, the flame front reaches the absorption path indicated by the fast drop of the fuel density. Remaining fuel shows an incomplete combustion, because the fuel density is much higher than fuel density measured after an ideal combustion.

3.4 Approximation Method for Time Resolved Lambda Values

The measurements of the single engine cycles were performed on a test bench, which enables to determine an averaged lambda-value for multiple cycles by a lambda probe in the exhaust pipe. The data record of the absorption signals, at different preset lambda values started when a stable lambda value was measured by the lambda probe. With the knowledge of the air/fuel-ratio (lambda-value) and the pressure p_{cyl} as well as the adiabatic temperature T_{cyl} it is possible to calculate the stoichiometric amount of air in the cylinder by using Eq. 3, 5 and 6 at the time the intake valves close. By assuming a homogenous distribution of air and an adiabatic compression until ignition, the time resolved density of air ϱ_{air} can be calculated with Eq. 10. The fuel density ϱ_{fuel} was determined by the measured transmissivity and the calibration function presented in chapter 3.2.

Eq. 11 shows the combination of the time resolved densities of air ϱ_{air} and fuel ϱ_{fuel} to an air/fuel-ratio. In reference to the stoichiometric air/fuel-ratio, which is given with m_{air}/m_{fuel}= 14.7, the lambda value can be calculated.

$$\lambda = \frac{m_{air}/m_{fuel}}{\left(m_{air}/m_{fuel}\right)_{sto}} \cdot \left(\frac{V_{cyl}/V_{cyl}}{1}\right) = \frac{\rho_{air}/\rho_{fuel}}{\left(m_{air}/m_{fuel}\right)_{sto}} = \frac{\rho_{air}/\rho_{fuel}}{14.7} \tag{11}$$

3.5 Time Resolved Air/Fuel Ratio Measurements

The three crank angle resolved fuel density variations (misfire, ideal combustion and delayed combustion), presented in chapter 3.3, will be used in combination with the theoretical determined air density to calculate the crank angle resolved lambda values with the approximation method of chapter 3.4.

Fig. 12 shows the crank angle resolved development of the lambda values during the homogenization phase just before ignition. There is a homogenous air/fuel-mixture until the ignition because of the early fuel injection into the intake manifold, which is indicated by the almost constant lambda value. After ignition, there is a time delay of nearly 20 degree crank angle for the "ideal" combustion until the flame front reaches the measurement region. The rising lambda value indicates the lack of detectable hydrocarbons in the measurement path after the flame front has passed this area.

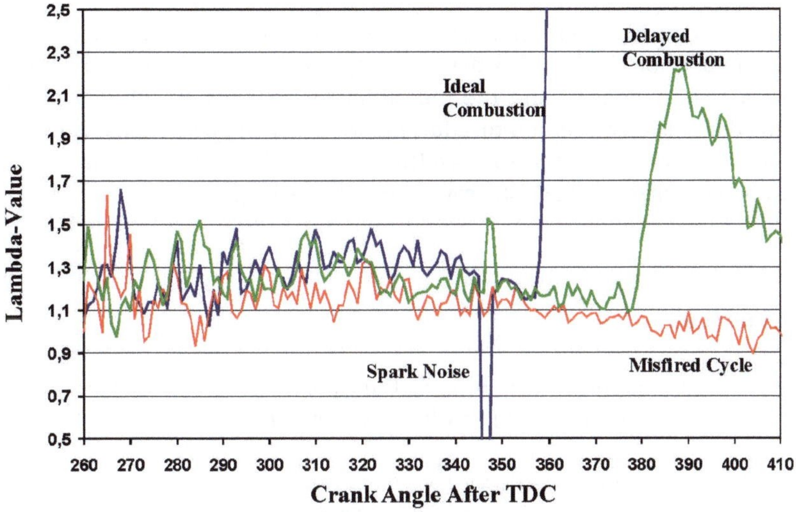

Fig. 12. Crank angle resolved lambda values for single engine cycles and a preset λ-value of 1.1

The time delay between the ignition and first detection of the combustion for the "delayed combustion-cycle" is about 40 degree crank angle, which is indicated by the rising lambda value at 380 degree crank angle. It is an incomplete combustion because there is still an amount of fuel present in the measurement path which leads to an air/fuel-ratio with a lambda value of about $\lambda = 2.0$. The misfired cycle shows a nearly constant lambda value of $\lambda = 1.1$ during the whole compression phase.

4 Summary and Conclusions

This publication describes the test of a fuel density sensor for automotive applications. The measurements were carried out on a SI production engine by using ordinary gasoline Euro Super.

The variations of fuel density during the intake and compression stroke were analyzed as well as the air/fuel-ratio dependent differences in the delay between ignition and combustion. A variation of preset lambda values showed that the system is able to resolve slightly different fuel densities during the homogenization phase.

Single engine cycles, with particular characteristics like misfire and delayed combustion, were investigated in this study and compared to the fuel density variations during ideal combustion. The time resolved lambda value was also determined for particular cycles by an approximation method which uses the calculated air density and the measured fuel density. Specific characteristic behavior of the air/fuel-mixture-formation at starting conditions and variations in the combustion processes resulting from air/fuel-ratio fluctuations during the homogenization could be revealed in this study.

References

[1] Zhao, F., Taketomi, M., Nishida, K., Hiroyasu, H., Quantitative Imaging of the Fuel Concentration in a SI Engine with Laser Rayleigh Scattering, SAE Paper 932641, 1993.

[2] Lee, K. H., Foster, D. E., Cycle-by-Cycle Variations in Combustion and Mixture Concentration in the Vicinity of Spark Plug Gap, SAE Paper 950814, 1995.

[3] Knapp, M., Luczak, A., Beushausen, V., Hentschel, W., Manz, P., Andresen, P., Polarization Separated Spatially Resolved Signal Laser Shot Multispecies Analysis in the Combustion Camber of a Realistic SI Engine with a Tuneable KrF Excimer Laser, Proc. Combust. Inst. 26:2589-2596, 1996.

[4] Miles, P. C., Dilligan, M., Quantitative In-Cylinder Measurements Using Broadband Spontaneous Raman Scattering, SEA Paper 960828, 1996.

[5] Egermann, J., Koebcke, W., Ipp, W., Leipertz, A., Investigation of the Mixture Formation inside a Gasoline Direct Injection Engine by Means of Linear Raman Spectroscopy, Proc. Combust. Inst. 28: 1145-1152, 2000.

[6] Baritaud, T. A., Heinze, T. A., Gasoline Distribution Measurements with PLIF in a SI Engine, SAE Paper 922355, 1992.

[7] Zhao, F., Taketomi, M., Nishida, K., Hiroyasu, H., PLIF Measurements of the Cyclic Variation of Mixture Concentration in a SI Engine, SAE Paper, 940988, 1994.

[8] Berckmüller, M., Tait, N., Lockett, R., Greenhalgh, D., Ishii, K., Urata, Y., Umiyama, H., Yoshida, K., In-Cylinder Crank-Angle-Resolved Imaging of Fuel Concentration in a Firing Spark-Ignition Engine Using Planar Laser-Induced Fluorescence, Proc. Combust. Inst. 25:151-156, 1994.

[9] Reboux, J., Puechberty, D., Dionnet, F., A New Approach of Planar Laser Induced Fluorescence Applied to Fuel/Air Ratio Measurements in the Compression Stroke of a S.I. Engine, SAE Paper 941988, 1994.

[10] Collings, N., A new technique for measuring HC Concentrations in Real Time, in a Running Engine, SAE Paper 880517, 1988.

[11] Cheng, W. K., Galliot, F., Collings, N., On the Time Delay in Continuous In-Cylinder Sampling from IC Engine, SAE Paper 890579, 1989.

[12] Galliot, F., Cheng, W. K., Cheng, C., Sztenderrowicz, M, Heywood, J. B., and Collings, N., In-cylinder Measurements of Residual Gas Concentrations in a Spark Ignition Engine, SAE Paper 900485, 1990.

[13] Alger, T., Hall, M. J., Matthews, R., Effects of In-Cylinder Flow on Fuel Concentration at the Spark Plug, Engine Performance in a DISI Engine, SAE Paper 2002-01-0831, 2002.

[14] Alger, T., Hall, M. J., Matthews, R., Fuel Spray Dynamics and Fuel Vapor Concentration Near the Spark Plug in a Direct-Injected 4-Valve SI Engine, SAE Paper 1998-01-0497, 1998.

[15] Koenig, M., Hall, M. J., Cycle-Resolved Measurements of Pre-Combustion Fuel Concentration Near the Spark Plug in a Gasoline SI Engine, SAE Paper 981053, 1998.

[16] Koenig, M., Hall, M. J., Measurements of Local In-Cylinder Fuel Concentration Fluctuation in a Firing SI Engine, SAE Paper 971644, 1997.

[17] Koenig, M., Stanglmaier, R., Hall, M. J., Matthews, R., Mixture Preparation During Cranking in a Port-Injected 4-Valve SI Engine, SAE Paper 972982, 1997.

[18] Hall, M., Matthews, R., Mid-IR Fiber Optic Sensors for internal Combustion Engines, 5th International Symposium on Internal Combustion Diagnostics, Baden-Baden, 2002.

[19] Tomita, E., Kawahara, N., Shigenaga, M., Nishiyama, A. Dribble, R. W., In situ measurement of hydrocarbon fuel concentration near a spark plug in an engine cylinder using the 3.392 µm infrared laser absorption method: discussion of applicability with a homogeneous methane–air mixture Meas. Sci. Technol. 14, 1350-1356, 2003.

[20] Nishiyama, A., Kawahara, N., Tomita, E., Fujiwara, M., Ishikawa, N., Kamei, K., Nagashima, K., In-Situ Fuel Concentration Measurement Near Spark Plug by 3.392 µm Infrared Absorption Method – Application to a Port Injected Lean-Burn Engine, SAE Paper 2004-01-1353, 2004.

[21] Nishiyama, A., Kawahara, N., Tomita, E., In-Situ Fuel Concentration Measurements Near Spark Plug by 3.392 µm Infrared Absorption Method – Application to Spark Ignition Engine, SAE Paper 2003-01-1109, 2003.

[22] Tomita, E., Kawahara, N., Nishiyama, A., Shigenaga, M., In situ measurement of hydrocarbon fuel concentration near a spark plug in an engine cylinder using the 3.392 µm infrared laser absorption method: application to an actual engine Meas. Sci. Technol. 14, 1357-1363, 2003.

[23] Thiele, O., Fiberoptical analysis of the mixture formation process in gasoline direct injection combustion engines, PhD Thesis, University of Göttingen, 2004 http://webdoc.sub.gwdg.de/diss/2005/thiele/thiele.pdf.

[24] Grosch, A., Beushausen, V., Thiele, O., Grzeszik R., Crank Angle Resolved Determination of Fuel Concentration and Air/Fuel Ratio in a SI-Internal Combustion Engine Using a Modified Optical Spark Plug, SAE Paper 2007-01-0644, 2007.

Alexander Grosch, Volker Beushausen
Laser-Laboratorium Göttingen e.V.
Hans-Adolf-Krebs-Weg 1
D-37077 Göttingen
Germany
alexander.grosch@llg-ev.de
volker.beushausen@llg-ev.de
www.llg-ev.de

Olav Thiele
La Vision GmbH
Ann-Vandenhoeck-Ring 19
37081 Göttingen
Germany
othiele@lavision.de

Keywords: infrared, IR, absorption, spark-plug-sensor, fiber optic, fuel concentration, air/fuel-ratio, combustion engine

Two-Line LIF-Emission Thermometry for Gas-Temperature Determination in IC-Engines

R. Müller, V. Beushausen, Laser-Laboratorium Göttingen e.V.

Abstract

It is still a huge challenge in the combustion diagnostics to determine the temperature of the fuel/air-mixture within a combustion chamber. But the knowledge of the temperature is a essential information for developing engine systems. In this work a measurement technique is introduced, which could be a solution for this problem. This technique is based on two-line-emission-spectroscopy in which formaldehyde is used as a tracer molecule. The fluorescence excitation is generated by a pulsed UV-laser with a wavelength of 355nm. A chamber is used, which enables to vary the temperature and the pressure. Measurements are done under motor-driven conditions. Thereby the fluorescence behaviour of the tracer substance is investigated both in air environments and in fuel/air-mixtures. The extracted results show a direct temperature-dependent relation of the detected fluorescence signals. These conclusions establish the analytical basis for a measuring system to determine the temperature of fuel/air-mixtures within combustion engines.

1 Introduction

Applications of internal combustion engines are found all over the world, especially in cars and commercial vehicles. Therefore it is of paramount importance to minimize the entire pollutant emission and consumption of the slender fossil fuel resources, which are caused by this technique. For developing engines this means to decrease the fuel consumption and to increase the degree of efficiency concurrently. In order to achieve success in these developments, an accurate knowledge of the physical and chemical processes within the combustion chamber is necessary. For this purpose mixture formation, ignition and combustion need to be to examined. In this connection one important parameter is the temperature of the fuel/air mixture, which has a crucial influence. In particular this is of great importance for the realization of the HCCI-combustion concept (HCCI: Homogeneous Charge Compression Ignition). Here auto-ignition of the fuel/air mixture is generated due to compression induced temperature increase.

An acquisition of this physical parameter with high spatial and temporal resolution is still a huge challenge. For that a contactless optical method is most suitable. Some techniques were developed in the past. Schulz et al. [1] used Rayleigh scattering to determine the temperature. This method has the disadvantage, that the intensity of the scattered laser light is very weak. Because of scattering light caused by the combustion chamber wall, fuel droplets and sooty particles strong background signal is generated, which produces a low signal-to-noise-ratio. This makes an application difficult. Also Raman spectroscopy was done by Zur Loye et al. [2] in order to determine the temperature. Again a weak intensity of the signal presents a problem and also the small gaps of the Stokes and Antistokes lines demand a very high spectral resolution. Laser-induced fluorescence (LIF) of NO and OH radicals were deployed for temperature measurements. Palmer et al. [3] observed NO radicals in a shock tube experiment and Arnold et al. [4] examined OH radicals in flames. Both species are present only during the combustion process. Therefore this method can not be used to determine the temperature of the fuel/air mixture before ignition. For temperature determination based on simultaneous scanning of the distribution of two energy levels in the electronic ground state, usually two different excitation lasers are necessary. Hence the realization needs a complex set-up. Einecke et al. [5] used it for examining 3-pentanone and got a good verification of the temperature. Contrary to the common LIF-methods, which detect the thermally excited population distribution in the electronic ground state, this work utilizes a two-line fluorescence method, which is based on detecting the vibronic population distribution of the tracer molecule formaldehyde in the electronically excited state.

2 Measurement Method

In this LIF-method formaldehyde is used as a tracer molecule. In the fluorescence spectrum of formaldehyde the vibronic transitions can be clearly detected. Furthermore Shibuya et al. [6] showed, that a thermal population distribution of the vibronic levels caused by collision-induced redistribution is finished within some picoseconds. The fluorescence lifetime of formaldehyde is quite longer so that thermalization is completed early during fluorescence emission. The population of the energy levels can be described via the Boltzmann distribution. By simultaneous probing the population of two thermalized electronically excited vibronic states by detecting two fluorescence emission bands, the vibronic temperature of the tracer molecules can be determined.

Usually two-line temperature measurements are based on simultaneous probing of the populations of two vibronic levels in the electronic ground state of molecules. In that case two different excitation lasers are needed. In this work a method is used, which analyzes emission spectra of electronically excited states. This has the advantage, that only one excitation laser is needed. Therefore the technical complexity is strongly diminished.

The applied two-line fluorescence method uses the emission of the electronically excited formaldehyde molecule, in order to get the temperature information. Here three emission lines are detected simultaneously. Via the Arrhenius equation the intensity-ratio of two emission lines is connected to the vibrational temperature of the tracer molecules. Thus a calibration function can be derived, which establishes a relationship between fluorescence intensity ratio and temperature.

The aim of this work is a basic examination of the utilizability of the described measurement technique for the evaluation of the gas temperature during mixture formation and compression inside the cylinder of Otto-engines. Therefore the parameters gas pressure, gas temperature and gas composition were varied systematically. It was evaluated, if cross sensitivites of pressure and gas composition variations to the measured temperature occur.

3 Measuring System

In order to generate static motor-driven conditions, a temperature and pressure controlled gas bomb with three fused silica windows is used (Fig. 1). This chamber can be evacuated and is compression-proof up to 30bar. In the chamber a gas temperature of 200°C can be reached. Because under normal conditions formaldehyde is existent in the polymerized form para-formaldehyde, the gas bomb and all tubes are heated up to 120°C before filling. Para-formaldehyde breaks down to gaseous formaldehyde at around 80°C. Therefore re-polymerization of formaldehyde and hence an alteration of the previously defined formaldehyde concentration is avoided. Afterwards synthetic air was filled into the chamber by a heated choke valve and mixed up with formaldehyde homogeneously. Several mixture ratios were examined. During the filling with the components the temperature of the gas bomb was kept to approximately 90°C. After generation of the targeted gas composition and reaching thermal equilibrium, the chamber was slowly heated up to 200°C. During the heat-up phase the formaldehyde was excited by a pulsed frequency-tripled Nd: YAG laser (Quantel Brilliant B, wavelength 355 nm, repetition rate 10 Hz). The laser beam diameter was 7 mm and the laser energy was 60 mJ. Perpendicular to the laser beam the fluorescence was detected. With the help of a mirror and a biconvex lens (f= 100 mm) the fluorescence light was focused onto a slit of a spectrometer (Oriel Multispec, focal length 125 mm, Grating 2400 L/mm). The spectrometer is coupled to an image-intensified CCD-camera (LaVision, Flame Star 2), in order to acquire the spectra during the heat-up phase continuously. An optical filter (Schott GG 375) is used in order to suppress elastically scattered laser light. A thermocouple was used to determine the temperature within the gas bomb simultaneously to the spectroscopic measurements.

Measurements were done with mixtures of formaldehyde and synthetic air at different concentrations and pressures. Furthermore a formaldehyde enriched isooctane/air mixture was examined. For establishing this mixture primarily para-formaldehyde was solved in ethanol at a temperature of 90°C. This formaldehyde/ethanol solution can be easily mixed up with isooctane. In order to get a defined λ-value inside the gas bomb the required amount of fuel/formaldehyde mixture was injected through a septum by a syringe.

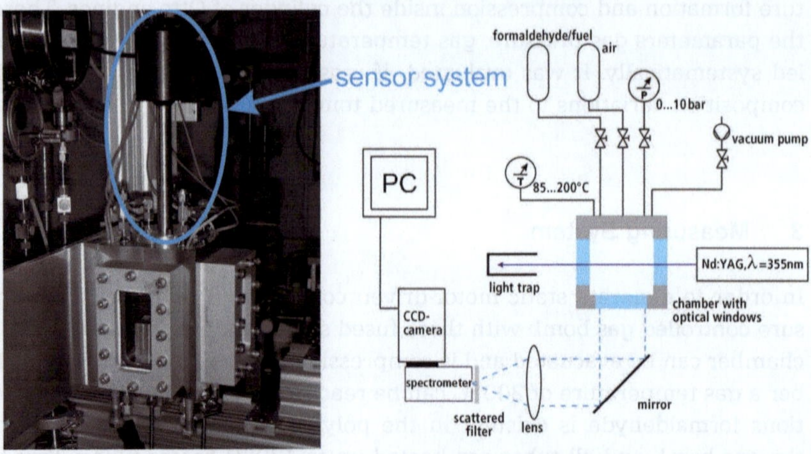

Fig. 1. Left: Temperature and pressure controlled gas bomb, here with the sensor system described in chapter 4. Right: Scheme of the experimental set-up.

3.1 Results

The applied method was adopted from Burkert et al. [7] to investigate temperature fields of flames under normal conditions. The work described here in particular deals with the extension of the applicability of the method to higher pressures and reactive compounds in a moderate temperature range.

Also there was a substantial interest in finding two fluorescence lines, whose intensity ratio does not exhibit interferences from pressure and mixture composition variations. The acquired spectral range comprehends the transitions $2^0{}_1 4^0{}_1$ (395 nm), $4^0{}_3$ (405 nm) and $2^0{}_1 4^1{}_2$ (413 nm). The detected vibration spectrum is shown in Fig. 2. The curves are normalized to 413 nm in order to highlight the temperature induced intensity fluctuations. Calibration curves are obtained by plotting the line intensity ratios over the respective temperatures.

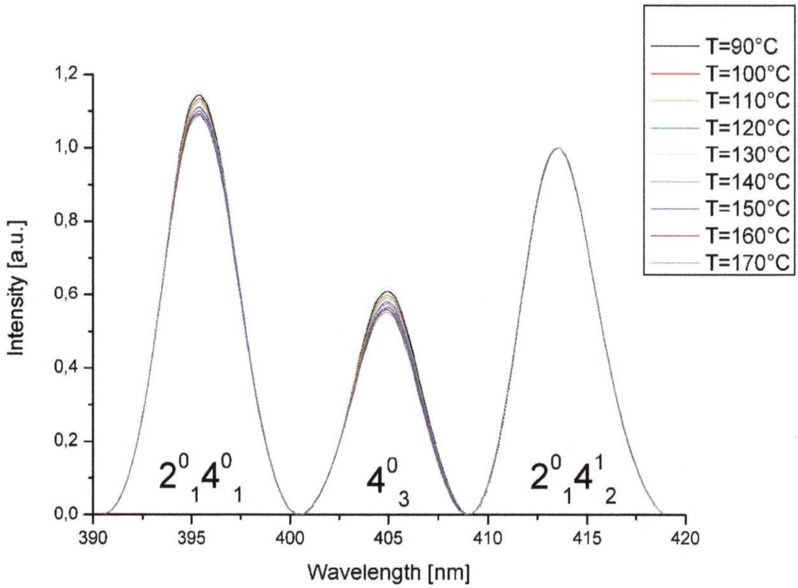

Fig. 2. Normalized fluorescence emission lines of formaldehyde.

Under engine conditions a mixture of fuel and air is generated. Therefore the oxygen induced fluorescence quenching of formaldehyde was examined. Thereby 100 mbar formaldehyde was mixed up with different amounts of synthetic air. The air fraction of the respective mixture was chosen, that total pressures from 1 bar up to 10 bar are derived. Ratios of all possible combinations of the detected fluorescence lines were calculated and it was checked, which ratio shows the strongest temperature dependency and the lowest cross sensitivity to pressure and composition variations. The analysis showed, that the line ratio 405 nm / 413 nm shows nearly an identical dependency for all observed cases (Fig. 3). The variation of the curves is fairly small and shows only minor pressure dependency. In addition, it appears that an alternation in concentration does not have any influence on the result, because in each measurement the amount of formaldehyde was constant, but the fraction of air increases proportional to the total pressure.

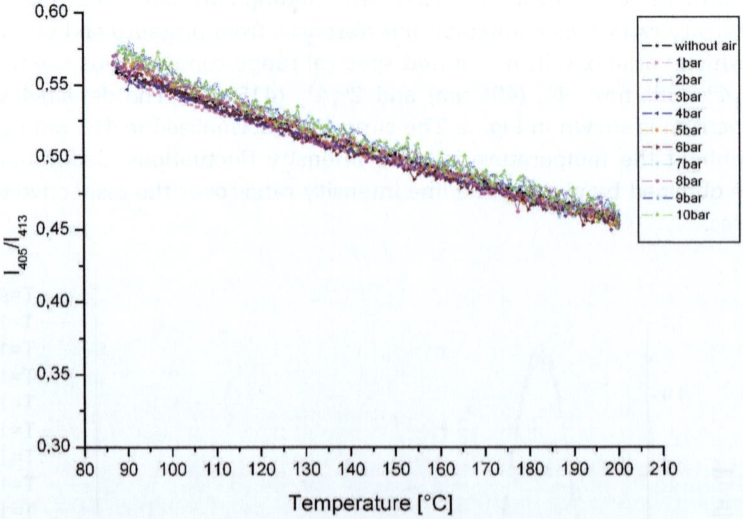

Fig. 3. Dependency of the formaldehyde fluorescence line ratio 405 nm / 413 nm on temperature for different air pressures.

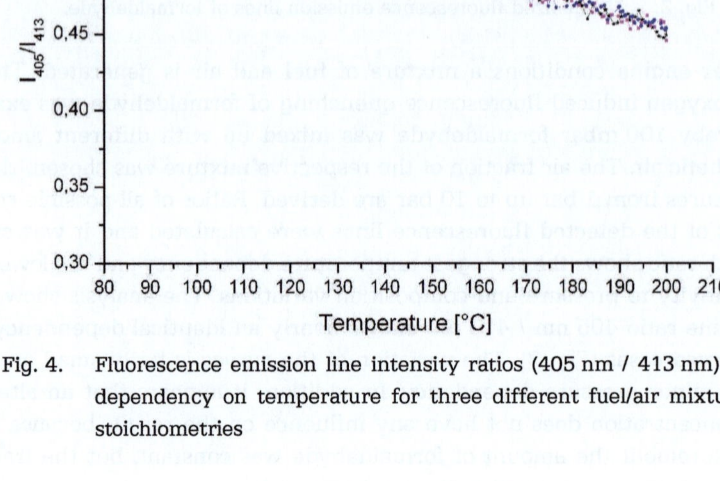

Fig. 4. Fluorescence emission line intensity ratios (405 nm / 413 nm) in dependency on temperature for three different fuel/air mixture stoichiometries

This behaviour still exists in case a fuel/formaldehyde/air mixture is used instead of the pure formaldehyde/air mixture. For these experiments in para-formaldehyde is previously solved in a small amount of ethanol and then mixed up with the model fuel isooctane. With this solution and with air different mixtures were generated inside the gas bomb. Measurements were done for defined λ-values. A lean mixture ($\lambda = 1{,}7$), a rich mixture ($\lambda = 0{,}6$) and an almost stoichiometric mixture ($\lambda = 1{,}1$) were examined. All three samples show a nearly identical behaviour (see Fig 4).

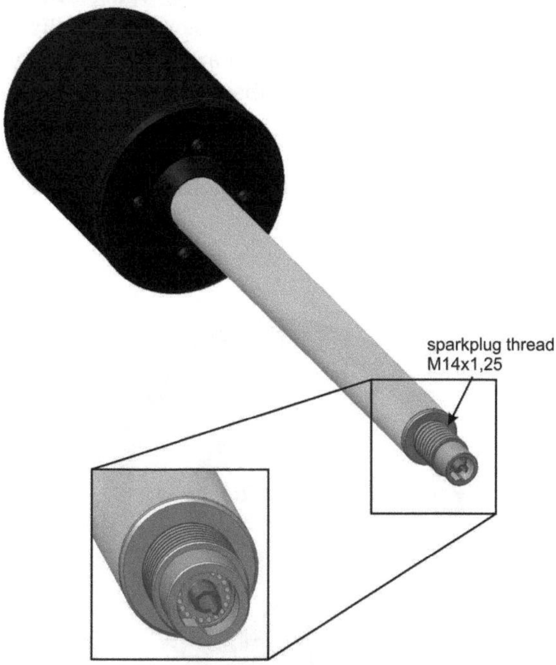

Fig. 5. The developed sensor system can be installed into an engine instead of a sparkplug.

4 Fiber-optical Sensor System

A fiber-optical sensor system was developed, which enables LIF-measurements in production engines without the usually necessary modifications of the engine geometry. The optical access is generated by constructing a fiber-optical sensor which fits into the sparkplug thread of the engine (Fig. 5). This allows conservation of the thermodynamical properties of the engine, because no other window needs to be established in the cylinder wall or the piston.

Instead of a sparkplug the sensor can be placed into the combustion chamber wall so that measurements under unfired engine conditions can be accomplished. In this concept the laser light is coupled into the combustion chamber by a fused silica rod, which is axially arranged in the sensor and extends into the measurement volume (see Fig. 6). This rod has a 45° polished end face, which makes sure that the light is coupled out normal to the sensor axis until it is reflected by the circular metal surface which is the outer rim of the sensor head. Due to multiple reflections complete illumination of the ring-shaped detection volume should be achieved. The gas, which is inside this volume, will be excited and exhibits fluorescence emission. Fused silica fibers are used to detect the fluorescence light. These are arranged circularly around the guidance of the fused silica rod. These fibers guide the fluorescence light to the spectrometer where it is analyzed.

First performance tests were carried out within the temperature and pressure controlled gas bomb.

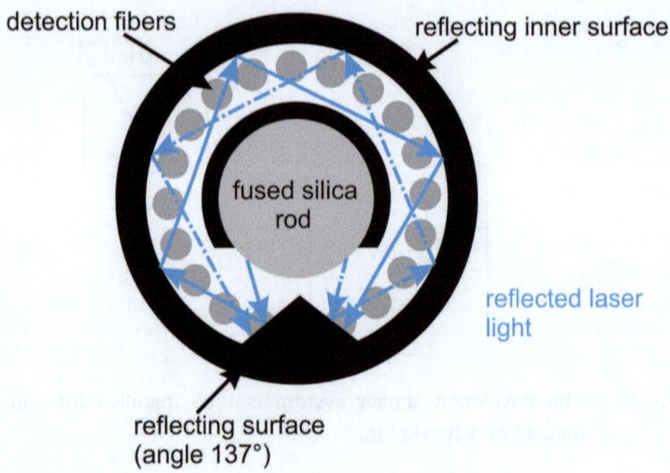

Fig. 6. Optical path of the laser light in the sensor system.

4.1 Results

A formaldehyde/air mixture (100 mbar formaldehyde and 900 mbar synthetic air) was filled into the chamber with a total pressure of 1 bar. During the tests the gas bomb was heated up continuously from 90°C to 150°C and spectroscopic measurements were carried out simultaneously. Unlike an internal combustion engine the walls of the gas bomb, where the sensor head is screwed in, is not cooled (engine wall temperature: ~90 °C). Therefore it is very likely that the complete front part of the sensor head will heat up to the gas temperature inside the bomb. Due to the limited temperature stability of the adhesives which were used inside the sensor, the final gas temperature was restricted to 150 °C.

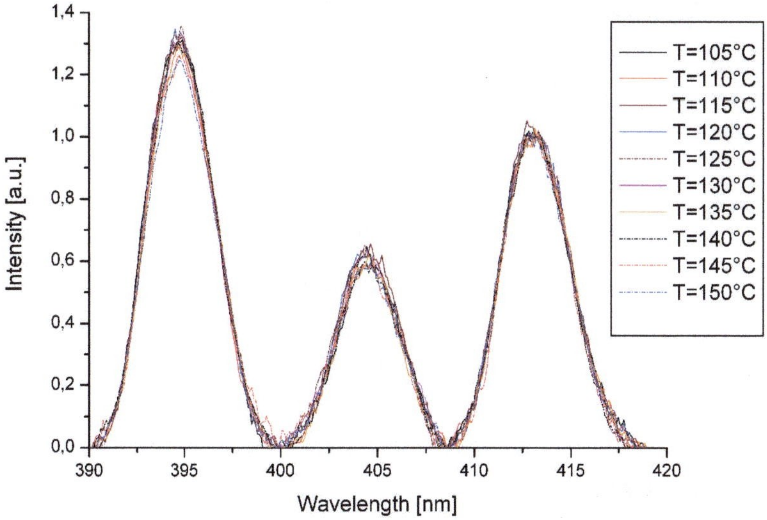

Fig. 7. Normalized fluorescence emission lines measured with the sensor system.

Although only low fluorescence intensity could be detected with the new setup the first tests of the sensor show promising results (see Fig. 7). The three vibrational transitions are clearly detectable but much stronger noise is overlapping the signal. It turned out, that the bad signal intensity is due to laser induced destruction of the roof shaped mirror surface which is hit by laser light (see Fig. 6). Nevertheless despite the very low fluorescence signal detected the temperature dependency of the line ratio 405 nm / 413 nm could be reproduced (see Fig. 8). Improvements are underway to avoid laser induced damaging of the mirror surfaces.

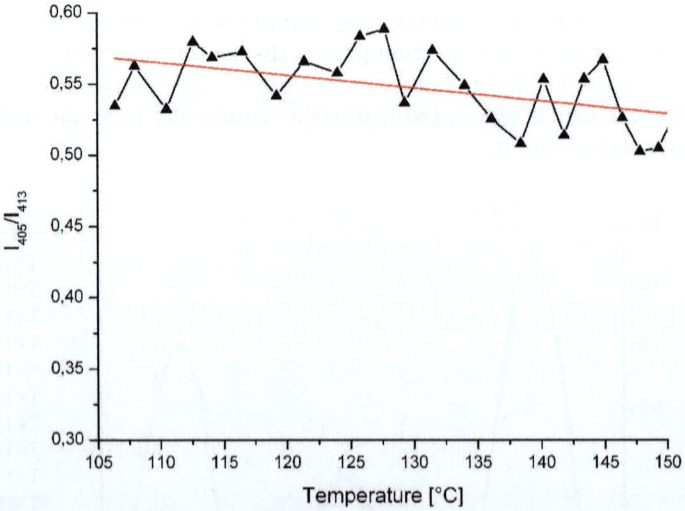

Fig. 8. Dependency of the line ratio 405 nm / 413 nm on gas temperature, measured by the newly developed sensor system (first results).

5 Summary

In this work an optical measurement method was presented, which can be used to determine the temperature of fuel/air mixtures inside combustion engines. The method is based on the temperature dependency of the intensity ratio of fluorescence emission bands of the tracer substance formaldehyde. Experiments were carried out inside a temperature and pressure controlled gas bomb where stable engine like conditions can be generated. Different temperatures, pressures and fuel/air mixtures were investigated in order to evaluate the applicability of the technique for determination of the gas temperature inside the cylinder of spark ignition engines. The results are very promising and display that gas temperatures can be measured without strong influence of varying gas-pressure and fuel/air ratios.

Since the technique shall also be applied to production engines with limited optical access a fiber optic sensor system is constructed which can be screwed into the spark plug thread of an engine head in order to determine the gas temperature at the spark plug position. First test show promising results.

References

[1] C. Schulz, V. Sick, J. Wolfrum,V. Drewes, M. Zahn, R. Maly: Quantitative 2D single-shot imaging of NO concentrations and temperatures in a transparent SI engine, 26th Symp. (Int.) on Combustion, the Combustion Institute, Pittsburgh, 1996, S. 2597.

[2] A. Zur Loye, D.A. Santavicca: Temperature and concentration measurements in an internal combustion engine using laser Raman spectroscopy, AIAA 18th Thermophysics Conf., 1983, Paper AIAA-83-1551.

[3] J.L. Palmer, B.K. McMillan, R.K. Hanson: Multi-line fluorescence imaging of the rotational temperature field in a shock-tunnel free jet, 1996, Appl. Phys. 63, S. 167.

[4] A. Arnold, B. Lange, T. Bouché, T. Heitzmann, G. Schiff, W. Ketterle, P. Monkhouse, J. Wolfrum: Absolute temperature fields in flames by 2D-LIF of OH using excimer lasers and CARS spectroscopy, 1992, Ber. Bunsenges. Phys. Chem. 96, S. 1388.

[5] S. Einecke, C. Schulz, V. Sick: Measurements of temperature, fuel concentration and equivalence ratio fields using tracer LIF in IC engine combustion, 2000, Appl. Phys. B 71, S. 717-723.

[6] K. Shibuya, E.K.C. Lee: Vibrational and electronic energy transfers from a single vibronic level of H_2CO, the 41 level, 1978, J. Chem. Phys. 69 (2), S. 758-766.

[7] A. Burkert, W. Triebel: Temperaturmessungen bis 800 K in Modellapparaturen und Verbrennungsmotoren basierend auf Laser-induzierter Fluoreszenz von Formaldehyd, Tagung Motorische Verbrennung, 2003, Berichte zur Energie- und Verfahrenstechnik, Heft 3.1, S. 263-271.

Ralf Müller, Volker Beushausen
Laser-Laboratorium Göttingen e.V.
Hans-Adolf-Krebs-Weg 1
37077 Göttingen
Germany
Ralf.Mueller@llg-ev.de
volker.beushausen@llg-ev.de

Keywords: fluorescence emission spectroscopy, gas-temperature, emission line ratio, formaldehyde, spark ignition engine

Comfort and HMI

Human Vehicle Interaction Based On Electric Field Sensing

C. A. Pickering, Jaguar Cars Ltd.

Abstract

The intervention of a human body part entering the path between a low frequency transmit electrode and a receive electrode causes a change in the displaced current measured at the receive electrode. This method of human interaction with low frequency electric fields can be used to create a wide range of interactive applications including human computer interfaces, interactive surfaces, musical instruments, virtual reality and automotive applications. This paper presents a review of electric field sensing techniques and applications involving Human Vehicle Interaction (HVI). The HVI applications are split into three categories: proximity sensing and touch detection; gesture recognition; user discrimination.

1 Introduction

It is possible to use multiple electrodes to create electric fields, and then measure the induced potentials and displacement currents caused by the proximity of a human body part. The term Electric Field Sensing (EFS) will be used to refer to a family of non-contact measurements of the human body that may be made with slowly varying electric fields [1]. These measurements can be used to measure the distance of a human hand or other body part from an object; this facilitates a vast range of applications for a wide range of industries. The unique benefits offered by low frequency electric field sensors include; no requirement for line of sight, fast response times, low power consumption, high resolution and they are safe to use.

2 Previous Work

One of the earliest forms of Electric Field Sensing was the non-contact musical instrument called the Theremin, invented in 1919 by a Russian Physicist named Lev Sergeyevich Termen (later changed to Leon Theremin).

Between 1919 and the early 1990's, few applications of electric field sensing have been found, mainly due to the lack of electronic devices that can be controlled by very tiny currents, then in the early 1990's the Massachusetts Institute of Technology (MIT) Physics and Media Laboratory carried out substantial research concerning media, musical and human computer interface applications of EFS [2].

This paper expands on the early work carried out by MIT and applies EFS theory and techniques to the automotive environment, this creates new technical challenges which are discussed in section 4.1. Before solutions to these automotive specific technical challenges can be discussed, it is first necessary to understand the basic theory of EFS.

3 Electric Field Sensing Modes of Operation

Fig. 1 shows the basic operation of Electric Field Sensing, this is a model that describes all 3 sensing modes for a single transmit-receive pair with a single target object [3]: Human Shunt Mode, Transmitter Loading Mode and Human Transmit Mode. The latter is used in HVI applications and is briefly described in section 3.1.

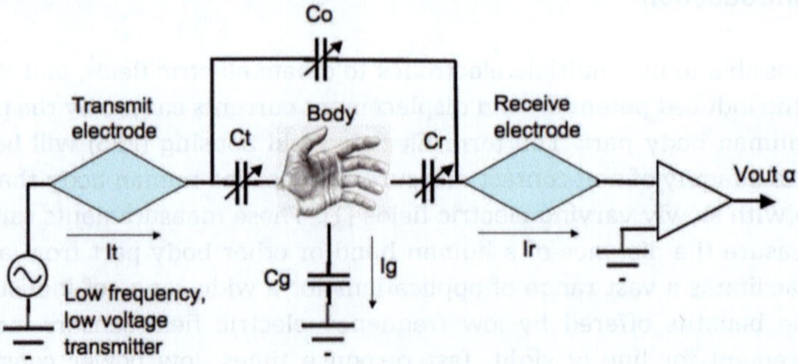

Fig. 1. Equivalent circuit for all modes of EFS

This simple hand sensor consists of two electrodes: a transmitter driven by a low frequency, low voltage signal, and a receiver that detects the transmitted signal through the capacitive paths given in Fig. 1. In order to reduce interference from ambient electromagnetic background, the receiver usually has a narrowband response centred at the transmit frequency, generally provided by a synchronous detection scheme [4].

The transmit and receive electrodes can be used in a variety of ways, each of which modifies these capacitances differently. These changing capacitances are seen as a changing current arriving at the receiver.

3.1 Transmit Mode

In transmit mode, the transmit electrode is put in contact with the users body, which then becomes the transmitter, either because of direct electrical connection, or capacitive coupling through the clothes, which is shown as path C_t which dominates.

When the hand moves, the spacing to the receiver changes, which changes the value of C_r. When the spacing from the hand to the receiver, r, is large, the received signal is approximately $1/r^2$ because the hand acts like a point object and the field falls off as $1/r^2$. By Gauss's Law, the induced charge on the receiver also goes as $1/r^2$. Since the potentials on the electrodes are defined by the circuit, we know the capacitance to be $C=Q/V$, and the received current $I_R=2\pi CV$. When the hand is very close to the receiver, C_r (typically) has the geometry of a parallel plate capacitor, and the signal goes to $1/r$ [1]. The Transmit mode sensing technique works very well for tracking the motion of a user in contact with a transmitter, as the received signals are only a simple function of the distance between the body and receive electrode, limb position can be easily estimated [4]. The Transmit Mode sensing technique is one that has been overlooked until recently.

4 System Hardware

The capacitance and displacement currents for electric field sensing are of the order of picofarads (10^{-12} F) and nanoamps (10^{-9} A), requiring sophisticated detection strategies [2]. A synchronous detection circuit is used to detect the transmitted frequency and reject all others, acting as a narrow band-pass filter. The displacement current can be measured with approximately 10 or 12 bit accuracy. Small displacement currents require good shielding, however the capacitance of shielded coaxial cable is orders of magnitude greater than the capacitance between the electrodes. Placing a current amplifier at the receive electrode allows higher frequencies to be used, the typical frequency range for electric field sensing human computer interface applications is 50 KHz to 150 KHz.

The only power consumed by the transmitter is the energy required to charge the capacitance of the transmitter electrode to the oscillating voltage. In practice the transmitter power is less than a milliwatt. This allows the design of very low power systems with no radio interference.

4.1 Automotive EFS Technical Challenges

Automotive experiments carried out by the author have demonstrated that when the transmit electrode is placed inside the seat cushion and receive electrodes are placed inside various mechanical push button switches, there are several factors which affect the amplitude and phase of the received signal level. These include; distance between the user and the transmit electrode - typically due to varying thickness of clothing or change of seating position, water contact between transmit electrode and occupant, signal coupling between vehicle occupants, and the loading affects of nearby objects at ground potential.

To overcome these challenges a reference receive electrode was placed in the seat cushion using additional electronics and software, the signal from the reference receive electrode was used to measure the signal strength on the user and then dynamically introduce compensation adjustments for those cases when the signal on the user was degraded by being earthed or increased by being wet. This research and the resolution of these system limitations resulted in a patent application [5].

5 Automotive EFS Applications

Many non automotive EFS systems have been developed by MIT, and Mitsubishi Electric Research Laboratories have also developed a desktop EFS system called DiamondTouch, these are noted for completeness and serve to illustrate the wide range of possible applications to which EFS can be applied [6].

EFS can also be used for a variety of automotive applications, examples include: placement of sense electrodes in the seat to detect if a person is sat on a seat or not, and by adding more sense electrodes around the seat area it is possible to detect whether an occupant is sat out of position allowing the airbags to be disabled in the event of a crash; using sense electrodes placed in a roof liner it may be possible to detect the very small head movements called micro-nods to detect the first signs of driver drowsiness. However, this paper will focus on automotive EFS applications relating to HVI and these can be split into three separate categories; (i) proximity sensing and touch detection, (ii) gesture recognition and (iii) user discrimination.

5.1 Proximity Sensing and Touch Detection

Proximity sensing, including touch detection, is the simplest form of EFS HVI and in the example below the system comprises of 3 switches, when a hand approaches one of these three switches, a display in front of the driver is triggered and as the hand moves closer to the switch, the display graphics change colour to help guide the drivers hand to the required switch. This system, shown in Fig. 2. below, was given the name "Phantom" and was jointly developed between Jaguar Cars and Pi Technology, Cambridge.

Fig. 2. Phantom system

Fig. 3. illustrates how the graphics change to visually guide the driver as the hand approaches.

Similarly, proximity sensing could be used to trigger the illumination of touch sensitive switches, which would be unlit when no hand is detected and illuminated when an approaching hand is sensed. To improve usability, a small vibrator could be added to the switch surface to provide tactile feedback when the switch is touched.

Fig. 3. Sequence of graphics changing with hand position:
no hand present (upper left), hand proximity detected (upper right), hand approaching knob (middle left), hand almost touching centre knob (middle right), hand touching centre knob (lower left), hand adjusting centre knob (lower right).

5.2 Hand Gesture Recognition

The primary goal of the author's research into hand gesture recognition for automotive human vehicle interaction is to identify and evaluate possible applications and driver safety benefits. Most current research is focussed on gesture interfaces for menu based secondary control systems since these potentially offer the most safety benefits. It is likely that hand gestures will be used as a supplementary method of user control similar to voice and steering wheel controls.

Because one-handed gestures used in human-human conversation are limited in number, the author does not believe that either mapping different hand gestures to a complete in-vehicle device such as a radio, or mapping each different type of in-vehicle control type is feasible. The use of selective mapping to theme and function appears to offer more realistic practical possibilities and potentially greater safety benefits [7]. After considerable analysis of all the possible selective themes and functions, it was found that each set of natural hand gestures could fit into the following 5 application domain classifications described below:

5.2.1 Pre-emptive Hand Gestures

A pre-emptive natural hand gesture occurs when the hand is moving towards a specific control type or device and the detection of the hand approaching is used to pre-empt the drivers intent to operate a particular control.

Examples of such functions could include operation of the interior courtesy light, as the hand is detected approaching the light in the roof console, the light would switch on. If the hand is detected approaching the light again it would switch off, thus the hand movement to and from the device being controlled could be used as a pre-emptive gesture. If this basic technique of simply detecting hand proximity is used then clearly there are limited potential applications because many controls are located together to provide logical grouping for the driver. To pre-empt which control the driver requires even at a distance of several centimetres would be prone to error. Therefore, the target applications are those devices and controls that are well separated from other controls. In addition to the interior light, other possible candidate applications might include lighting the door pocket when a hand approaches, operation of motorised sun visors to automatically fold down or away when a hand approaching is detected and increasing the brightness of switch graphic illumination at night when an approaching hand is detected.

5.2.2 Function Associated Hand Gestures

Function associated gestures are those gestures that use the natural action of the arm/hand to associate or provide a cognitive link to the function being controlled.

For example, moving the arm in an angular sweep pivoted about the elbow in front of the windscreen could be used to signify that the driver wishes to switch on the windscreen wipers. Similarly, moving the driver's hand/arm downwards along the side of the door window pillar is associated with opening the door window. These gestures have an action that can be associated with a particular function.

5.2.3 Context Sensitive Hand Gestures

Context sensitive hand gestures are natural gestures that are used to respond to vehicle system prompts or automatic events.

Context sensitive gestures to indicate yes/no or accept/reject could be thumbs-up and thumbs-down or even a horizontal and vertical wipe. These could be used to answer or reject an incoming phone call, an incoming voice message or an incoming SMS text message. The same yes/no gestures could also be used to accept or reject prompts for automatic navigation re-routing if the advanced navigation system has been informed of an accident ahead. Similarly, if low fuel is detected, the system could ask the driver if he/she would like to be automatically routed to the nearest fuel station. It may also be possible to use one of the yes/no gestures for other functions, for example, if a previously heard Traffic Announcement (TA) comes on as the driver is listening to a music CD, the same no gesture could be used to stop the TA and return to the CD.

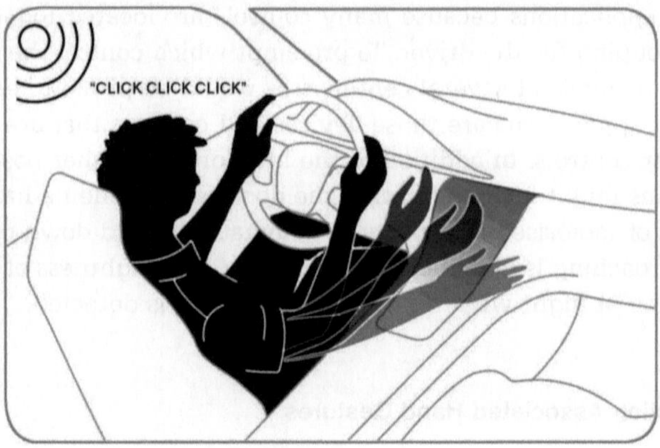

Fig. 4. Context sensitive accept gesture in response to incoming phone call

5.2.4 Global Shortcut Hand Gestures

Global shortcut hand gestures are in fact natural symbolic gestures that can be used at any time; the term natural refers to the use of natural hand gestures that are typically used in human to human communications.

It is expected that hand gestures will be selected whereby the user can easily link the gesture to the function being controlled. Possible applications could include fairly frequently used controls that present unwanted high visual workload, such as phone dial home, phone dial work or set navigation destination data entry to home.

5.2.5 Natural Dialogue Hand Gestures

Natural dialogue hand gestures utilise natural gestures as used in human to human communication to initiate a gesture dialogue with the vehicle, typically this would involve two gestures being used although only one gesture at any given time.

For example if the driver fanned his hand in front of his face, the system could interpret that the driver is too hot. An audible and/or visual prompt could be offered to the driver to ask if he/she would like all climate controls to maximum cold. If the driver then uses the context sensitive thumbs-up gesture, all the climate controls could be automatically set to low, including; switching on the air conditioning, setting the temperature to lowest setting, directing air vents at the drivers face and switching the heated/cooled seat to lowest temperature. Other dialogues could be initiated in a similar manner by using appropriate gestures, a natural gesture for I'm cold could initiate a dialogue of "do you want hottest settings?", or a natural gesture for I'm hungry or thirsty could be used to initiate a dialogue of "do you want to be directed to the nearest services or restaurant?". The applications are only limited by our ability to find a natural gesture to initiate the required meaningful dialogue; it is acknowledged that vision-based gesture systems are better suited to detecting natural dialogue hand gestures.

5.3 User Discrimination

User discrimination is achieved by using EFS in transmit mode as described in section 3.1 with separate transmit electrodes built into the driver and passenger seat cushions. A receiver electrode is built into each of the user controls. Under software control, the excitation signal is switched alternately to each vehicle occupant's transmitter electrode. For each occupant, while the excitation signal is present, the received signal level at each user control electrode is measured, as is the level on both occupants' seat mat receiver electrodes. These measurements are then processed by complex algorithms to provide the approximate impedance between each occupant and each of the controls.

These results may then be simplified to establish for each user control whether an individual user's hand is close, or touching it.

A signal indicating the identity of the user is then sent from the EFS module to a CPU that enables the authorised functions for that user, and activates a display through a graphics controller which provides visual feedback of the controller position and/or status to the user. For specific functions like alphanumeric text data entry, if the controller has sensed the driver, the display may not be activated since it may not be safe for the driver to carry out text entry when the vehicle is moving, however, if the passenger is detected, the controller could enable full functionality allowing the passenger to enter satellite navigation destination addresses when the vehicle is moving [8].

5.3.1 EFS User Discrimination HVI Applications

EFS is capable of hand proximity sensing, touch detection and gesture recognition. Using transmit mode, user discrimination can be applied to each of these. The following categories of HVI user discrimination features are possible: switch reduction; conditional passenger inhibit; driver enable; conditional driver inhibit; passenger enable.

A significant benefit of switch count reduction is improved driver safety. Presenting a lower number of switches to the driver causes less distraction to locate the required switch. This is achieved by using one common switch where currently two are used, for example dual zone temperature controls which provide duplicate temperature up and down switches for both the driver and passenger could be replaced by one set of switches. Similarly, by moving the powered window lift switches or seat control switches towards the centre of the vehicle, one set of switches could be used. In some applications the switches can be completely removed, for example the interior map reading light switches could be replaced by hidden electrodes to trigger the lights on proximity, and user discrimination could be added to allow the driver to switch both lights but allow the passenger to operate his/her map reading light only.

Applications of passenger inhibit and driver enable could include preventing the passenger from operating specific controls such as traction control, sports mode buttons, or possibly even electronic hand brakes and gear shifters that use the latest shift-by-wire technologies, however, the latter two applications may have complex legal and safety ramifications. For less critical systems it is conceivable that a configurable menu structure could be implemented to allow the driver to choose what controls the passenger can operate. Taken to the extreme, the passenger could be prevented from operating all climate control and in car entertainment functionality, which could be useful when driving with a child passenger!

For safety reasons, many automotive manufacturers currently disable specific vehicle functionality when the vehicle is in motion, and these can include alphanumeric text entry via touch screens for satellite navigation destination entry and SMS text messaging. With user discrimination it is possible to provide a more customer-friendly solution by allowing the passenger only to operate these features, this maintains driver safety benefits and at the same time enhances convenience. For example the driver will no longer need to stop to enter an address or send a text message via the touch screen as he can simply ask his passenger to do the task.

The prototype system hardware, which was developed for a vehicle demonstrator of user discrimination is now described. The donor vehicle was a Jaguar X-Type and user discrimination was implemented for the central rotary controller (Fig. 5), steering column pedal adjust switch (SCPA) and touch screen (Fig. 6).

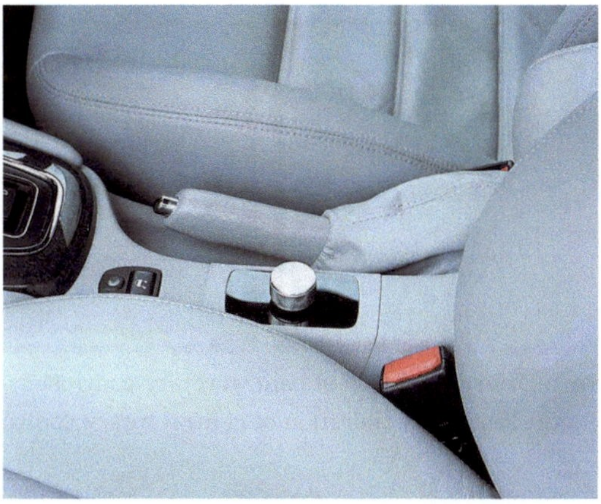

Fig. 5. Central rotary controller with user discrimination

Receive electrodes were designed and embedded in, or mounted on, each of these controls. An off-the-shelf display was mounted over the central air vents and used to indicate recognition of driver and/or passenger together with providing visual feedback when using the central rotary controller, see Figs. 7 and 8. Fig. 7 shows the display as seen by the driver when he operates the central rotary controller when the vehicle is moving, for safety reasons the navigation destination entry and mobile phone functions can be inhibited as shown. However, Fig. 8 shows that when the Passenger is operating the same central controller under the same vehicle conditions, full use of all functions including navigation and mobile phone functionality are available.

Fig. 6. Touch screen with user discrimination

Fig. 7. Left: Driver operation of central rotary controller

Fig. 8. Right: Passenger operation of central rotary controller

6 Limitations of Electric Field Sensing

There are several important limitations to electric field sensors in an automotive environment which are briefly discussed here: First, the electronics, typically capacitances of picofarads, create very tiny current signals at the receive electrodes. The supporting electronics must be carefully designed to provide adequate sensitivity while maintaining robust rejection of stray capacitance and electro-magnetic interference (EMI). Secondly, the automotive interior is limited by how much real estate is available for the sensor packaging. This directly affects sensing range and field shape [9].

Electric field sensors detect a bulk effect from the entire body in transmit or shunt mode. Therefore, care is needed in the design of input devices which are intended to measure isolated parts of the body, such as individual hands or fingers, so that the receivers will be responsive to movements of these, and the effect of the rest of the body can be offset by cross-referencing between sensor channels [10].

EFS in Transmit Mode for in-vehicle applications is also affected by numerous uncontrollable parameters including the size of user, earthed user, gloved hands, thickness of clothing, wet seats, wet user, wet receive electrodes and padding on seat.

7 Future Research and Conclusion

Further work will focus on system modelling and prototype development of automotive HVI and safety applications using electric field sensing in transmit mode. Further research and development will also be carried out to evaluate the scope of the identified limitations of electric field sensing in transmit mode for in-vehicle applications.

This Paper has presented an overview of HVI sensing techniques and applications for low frequency electric field sensors. In particular the human transmitter mode has been identified and described.

The benefits of electric field sensors were outlined; these include low power, low weight, high resolution and the ability to be aesthetically hidden since electric field sensors can penetrate non-conductive surfaces.

The limitations of transmit mode electric field sensors in an automotive environment were also briefly discussed.

The benefits and possible applications offered by electric field sensors are now being recognised with the emergence of new automotive applications such as occupant detection and classification, drowsiness detection, smart airbags and window safety systems. Since the original Theremin was developed in 1919, it has taken almost a century to get the necessary pre-requisite elements in place to ensure the successful and widespread use of EFS applications.

References

[1] Smith, J.R., Field Mice: Extracting Hand Geometry from Electric Field Measurements, IBM Systems Journal, Vol 35, Nos 3&4, 1996.

[2] Zimmerman, T., Smith, J.R., Paradiso, J.A., Allport, D. Gershenfeld, N., Applying Electric Field Sensing to Human-Computer Interfaces. In Proceedings of CHI '95 Denver CO, ACM Press, May 1995.

[4] Paradiso, J.A., Gershenfeld, N., Musical Applications of Electric Field Sensing, Draft 1.9, to be published in Computer Music Journal, April 1996.

[5] Pickering, C.A., Patent Application: Electric Field Sensing Proximity System. Patent Application: International Publication Number WO 2005/017727 A1. 2004.

[6] Pickering, C.A., Burnham, K.J., Richardson, M.J., A Review of Sensing Techniques and Applications with Low Frequency Electric Field Sensors: Proceedings 16th International Conference on Systems Engineering, ICSE 2003, Coventry University, 9-11 Sept. 2003.

[7] Pickering, C.A., Burnham, K.J., Richardson, M.J., A Research Study of Hand Gesture Recognition Technologies and Applications for Human Vehicle Interaction: Proceedings 3rd Institution of Engineering and Technology Conference on Automotive Electronics, Warwick University 28-29 June 2007.

[8] Pickering, C.A., Thorpe, I.R., Styles, J.M., Patent Application: Control Systems for a Vehicle Subsystem. International Publication Number WO 2004/022388 A1. 2004.

[9] Kithill, P.W., Capacitive Occupant Sensing, SAE Technical Paper Series, 982292, International Body Engineering Conference & Exposition, Detroit, Michigan, 29.9.1998.

[10] Allport, D., Zimmereman, T., Smith, J.R., Paradiso, J.A., Gershenfeld, N., Electric field Sensing and the 'Flying Fish', in the ACM-Springer Multimedia Systems Journal, special issue on Multimedia and Multisensory Virtual Worlds, Spring 1996.

Carl A. Pickering
Jaguar Cars Ltd.
Engineering Centre
Abbey Road
Whitley, Coventry CV3 4LF
United Kingdom
cpicker6@jaguar.com

Keywords: electric field sensing, proximity sensing, touch detection, gesture recognition, user discrimination

Networked Vehicle

Automotive 1 Gbit/s Link Goes Standard

T. Rothhaupt, INOVA Semiconductors GmbH

Abstract

The growth in complexity of highly integrated semiconductors is a challenge for the industry. Semiconductors with a lot of pins are bigger, more difficult to handle during manufacturing and are more vulnerable to malfunctions. With APIX (Automotive PIXel link) initially developed by INOVA Semiconductors there is now an industry accepted technology available to address these limitations. APIX is a 1 Gbit/s differential serial link capable of handling video streams in harsh automotive environment, enabling chip designers to replace parallel interfaces with a 2-pin serial interface for video pixel data reducing the number of pins on chips significantly. This addresses the need of next generation automotive applications using cameras and displays.

1 Introduction

Applications such as navigation, rear-seat entertainment, infotainment, rear view cameras, lane departure warning, and blind spot detection will increase the number of displays and cameras in the next generation of automobiles. Market intelligence is therefore forecasting the use of up to 8 cameras and up to 4 displays per car in the future. This means a lot of video data has to be created, processed and transferred between different electronic units in the car, requiring the latest advancements of semiconductor technology. The processing performance of handling several independent video streams even on the same chip is available. The critical issues to be considered are:

- Getting one or more video streams in and out of the same chip
- Keeping the size of the chip (in this case determined by the number of pins) as low as possible
- Having a light, thin, bidirectional and cheap physical connection between different ECUs (Electronic Control Units) bridging up to 15 m.
- Compatibility of solutions provided by different semiconductor manufacturers and guaranteed interoperatibility

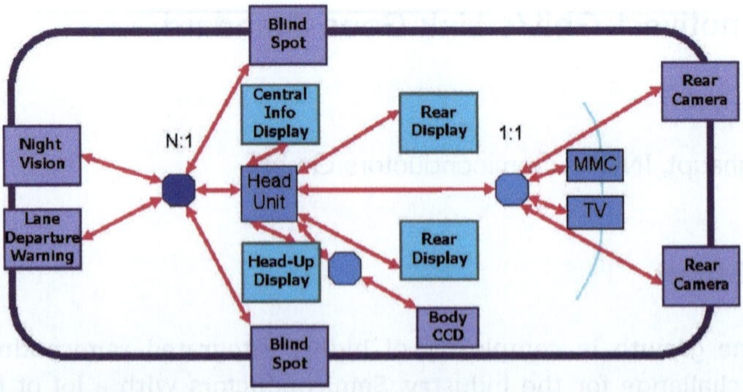

Fig. 1. Displays and cameras in cars (Inova)

APIX solves this issues and provides a technology available for integration.

2 APIX Technology

APIX is a long-distance high-speed bit serial link, DC-balanced with low latency and low EMI. It is optimized to connect central video/imaging processors systems to dispatched displays or cameras with only 2 wires.

2.1 APIX - Physical Interface

Selectable 0.5 or 1 Gbit/s operation, adjustable driver characteristics of the CML-IOs (Current Mode Logic) and integrated spread-spectrum clocking allow for the lowest EMI at maximum transmission distances. This significantly reduces the link's system costs while increasing the system functionality compared to simple serializer/de-serializer chip sets.

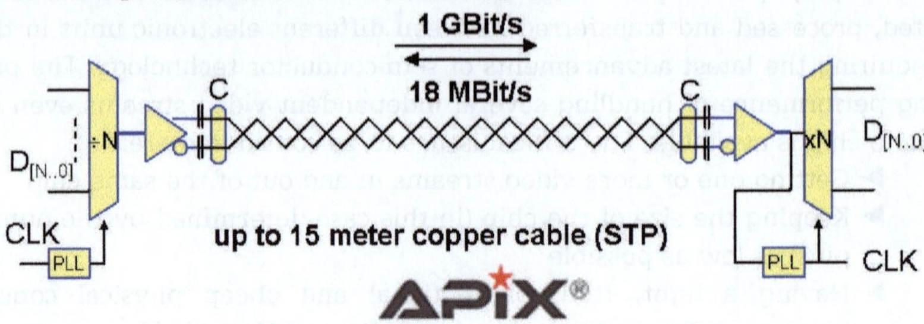

Fig. 2. APIX – physical principle (CML)

The link is configurable during power-up to achieve maximum link bandwidth at different interface widths supporting typical cameras and displays used in automotive applications. The individual drive current control and pre-emphasis adjustment, dynamic line coding and spread-spectrum clocking enable best-in-class EMI performance, especially in complex cable and connector assemblies with non matching constant impedances.

2.2 APIX - Capabilities

The available receiver and transmitter chipsets from INOVA (INAP125Txx and INAP125Rxx) offer a choice between two physical layer return paths: the embedded (Common Mode Jump Signalling on the video down stream link) or the separate return channel. These bi-directional sideband channels offer real-time operation and therefore are totally independent from the pixel clock fed into the video interface. The integrated bi-directional sideband channel offers an 18 Mbit/s capability that can be used, as an example, for I²C-compatible operations for controlling CMOS camera sensors or display settings. Even CAN-bus can be transferred almost transparently.

This allows the implementation of camera connections where the powering of the camera, transmission of the video data and controlling the camera remotely can be handled only by one pair of wires.

The 1 Gbit/s data rate allows the connection of state of the art TFT displays supporting popular resolutions. The devices are configurable to achieve maximum link bandwidth at different interface widths such as 10, 12, 18, 24 bits and of course transmission of typical control signals such as HSYNC, VSYNC and PxCLK. Again pixel data and control data can be send over one pair of wires in an optimized solution.

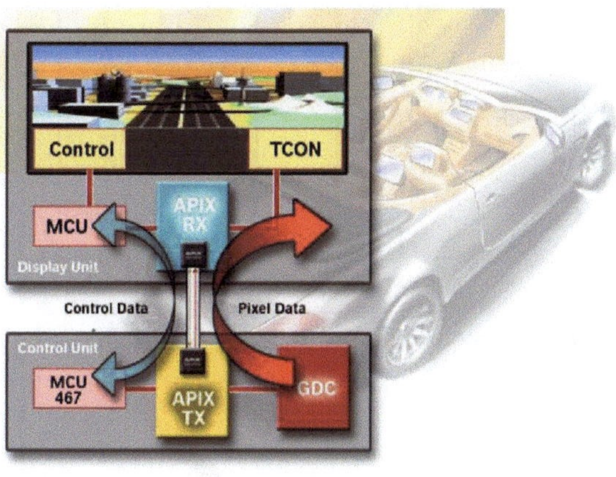

Fig. 3. APIX - display link (Inova/Fujitsu)

3 APIX - Automotive System Concepts

APIX physical interface and available discrete solutions are already used in different applications requiring high-speed data transfer over 2-wire copper cables. For applications enabling the system designer to use discrete or integrated APIX solutions from different vendors it is necessary to guarantee interoperatibility of the products available.

3.1 Interoperatibility

There are two aspects to be considered to achieve compatibility of APIX based systems:

- ▶ The physical interface
- ▶ A lower level protocol handling the data link

Every product with APIX technology available – discrete or integrated – is compatible with each other beyond the pure physical interface but also on a lower protocol level. This is optimized for automotive applications to ensure a stable link with error free data transfer covering areas such as CRC (Cyclic Redundancy Check) or ARQ (Automatic Repeat reQest) to name examples. Every semiconductor vendor integrating APIX products is bound to support the protocol. This is well documented and a prerequisite to do a proper APIX compatible design.

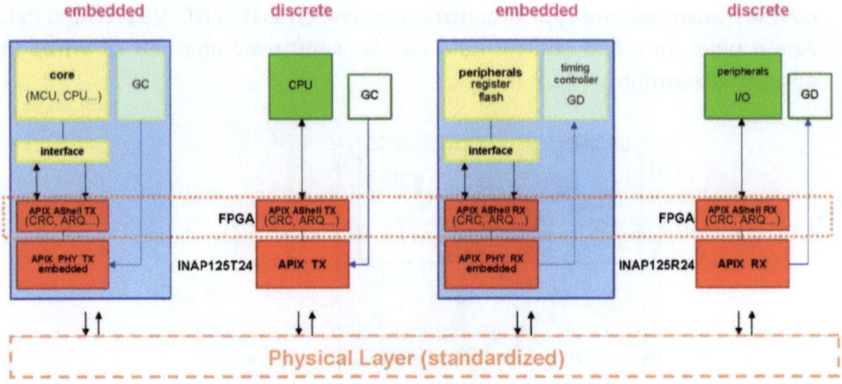

Fig. 4. APIX – system concepts (Inova)

3.2 Integration

Beyond the availability of discrete APIX receiver and transmitter chipsets, the technology is integrated in more complex VLSI chips such as graphic controllers/processors, imaging sensors, microcontrollers/processors or FPGAs.

Typical high end graphics controllers have several parallel interfaces for video capture (48 pins), PCI (44 pins), memory control (109 pins), and RGB (32 pins), thus increasing the number of pins and the size of chips significantly. Semiconductors with >250 pins and more are difficult to handle during production, less reliable and more expensive compared to lower pin count devices. The integration of more functional blocks is not limited by the size of the die or feature size of the semiconductor technology being used but by the number of pins and the size of the package. This is called pad limited.

By integrating APIX into more complex chips (such as graphic controllers) the number of pins can be reduced resulting in lower system cost due to the fact that now fewer chips can be used and boards can be designed smaller. On the other hand it also offers integration possibilities for other functionality.

Fujitsu and other semiconductor vendors have started to integrate APIX interfaces into the next generation of graphics controllers.

4 Validation

A key element for the acceptance and success of APIX as a solid and reliable technology for display and camera links is to ensure the interoperatibility between the different chips available. An independent reference to validate this is dedicated test equipment from Agilent.

	Features
Interfaces:	*4 digital in, 4 digital out pins* *APIX Rx Full speed/Half speed (*)* *APIX Rx Low speed (*)* *APIX Tx Full speed/Half speed (*)* *APIX Tx Low speed (*)*
PC Connectivity:	*100 BaseT Ethernet* *USB 2.0*
Temperature:	*Operating: -30°C to +70°C* *Storage: -40°C to +70°C*
Humidity:	*Operating: 15% - 95*
Safety standards:	*EN ISO/IEC 17025,* *IEC 61010-1/EN61010-1*
Power Supply:	*100…240 V AC with adapter in lab* *7…30 V DC*
Power Consumption:	*Less than 10 W*

The APIX protocol tester is covering the receiver and transmitter physical tests as well as the APIX lower level link protocol - including a 1 button compliance test, following the APIX specification.

By that system validation and interoperatibility can be guaranteed when using APIX implementations in different semiconductors.

Fig. 5. Agilent – APIX protocol tester (Agilent)

This of course can be used also during the manufacturing of APIX based electronic systems as well as regular car service cycles.

5 Conclusion

APIX provides an independent solution for linking cameras and displays in cars.

It allows connectivity between systems using APIX interface based semiconductors from different vendors, covering physical and link protocol integrity.

By design it is a reliable, serial high speed link, capable of transferring video pixel data and control information bidirectionally over one thin and light 2-wire copper cable, optimized for harsh automotive environments.

Interoperatibility, verification and test of APIX links is ensured by independent automated test equipment

APIX is available as standalone solution (INAP125xxx chipsets) or as an integrated solution initially at high pin count devices such a graphics controllers. There are more integration activities by semiconductor vendors on the way and it can be expected that there will be wide choice of specialized ICs available with APIX interfaces integrated.

References

[1] J. Valldorf, W. Gessner, Advanced Microsystems for Automotive Applications 2007. Springer, Berlin 2007.
[2] Fig. 3, Fig. 4, Fujitsu Semiconductors.
[3] APIX Data Sheet, INOVA Semiconductors.
[4] APIX Protocol Tester Product Document, Agilent.
[5] Fig. 5, Agilent.

Thomas Rothhaupt
Inova Semiconductors GmbH
Grafinger Str. 26
81671 München
Germany
trothhaupt@inova-semiconductors.de

Keywords: APIX, CML, current mode logic, Inova, Fujitsu, display, camera, camera link, display link, ATE

Components and Generic Sensor Technologies

Imaging Millimeter Wave Radar with Phased Array Antenna

R. Körber, Astyx GmbH
V. Ziegler, EADS Innovation Works
U. Schmid, Saarland University

Abstract

In the joint research project „RADARAUGE", funded by the Federal Ministry of Education and Research (BMBF), the development of a phased array radar sensor operated at 79 GHz with active MEMS-based micro-switches is targeted for automotive environmental recognition purposes. Basically, driver assistance systems need a narrow beam to detect objects at long distances and to reject reflections originating from objects located next to the road. To survey the traffic on several lanes in medium and short distances, however, a wide detection range is required. The movement of vehicles in the neighbourhood shall be monitored to support lane change manoeuvres etc. Both tasks can be covered by a phased array antenna, which provides electronic beam steering of a narrow lobe over a wide detection range. Beside the automotive application, the sensor can be used to detect wake-vortices of departing and landing aircrafts. This allows the optimization of the time intervals typically inserted between take-off and landing procedures.

1 Introduction and Technical Objectives

The overall technical goal of the project is the hardware realization and the corresponding final field test of a radar sensor with a compact design operated at 79 GHz. This device is mainly targeted at automotive applications, requesting a broad detection range ($\pm 40°$) at a simultaneously narrow lobe characteristics of each individual beam ($3° \times 5°$). These requirements are advantageously met using an antenna which is electronically steered. Due to this approach, it is possible to survey with the phased array radar sensor several lanes up to medium distances in the range of 30 to 80 m. Furthermore, a precise tracking of moving objects is feasible to determine the actual values for the velocity and for the position, respectively. Besides this usage, the device offers the qualification for a pre-warning function in the short range regime which enables a complete monitoring of the automobile up to 30 m. For the fabrication of the radar module with reduced volume being in addition competitive in the economical sense, a highly integrated assembly is targeted applying a combination of organic and ceramic materials, such as glass-fibre reinforced polyimide and LTCC (Low Temperature Co-fired Ceramics), in a most beneficial way. To meet the tough tolerances for the lateral dimensions of core components, such as the antenna elements, the applicability of thin film technology is of utmost importance. In addition, an electrical wiring approach is envisaged mainly based on thin film elements and hence, with a reduced number of wire bonds. In an accompanying paper [1], a novel process is developed to implement locally a tailored porosity in fired LTCC substrates in the surface-near region, thus reducing the permittivity substantially. For this procedure, phosphoric acid, a well established wet etchant in MEMS and microelectronic industry, is used. The phase control of the antenna elements is done by novel and low-cost RF-MEMS switching elements. In this context, high temperature stable materials for the movable structure in the RF micro switch, such as tungsten-titanium, are investigated [2]. Due to an enhanced temperature stability compared to aluminium-based alloys in the temperature range up to 400°C, a hermetic sealing of the cavity encapsulating this component seems to be possible applying e.g. a glass-frit based seal frame. A detailed characterization of the RF-based core component, especially in terms of reliability, accompanied by the corresponding modelling and simulation of the device performance ensures a product-oriented development approach, even at the early stage.

2 Device Layout and Pre-Evaluation of Core Components

2.1 Combination of Ceramic and Organic Materials

For a cost-effective approach guaranteeing simultaneously an optimized performance, the radar sensor consists in principal of a combination of organic and ceramic materials. Fig. 1 shows in a schematic view the corresponding layout of the front end. The TX/RX electronics as well as the matched corporate feed network for the different antenna lines are directly arranged on a low-loss multilayered LTCC substrate. The relatively high permittivity of the LTCC material (i.e. $\varepsilon_r = 7..8$) compared to air minimizes parasitic radiation effects arising from the feed network, thus resulting in an undisturbed antenna characteristics. When using LTCC technology, vias with a complex design and passive electronic components, such as resistors, capacitances and inductors, all needed for providing an appropriate DC supply to the RF components and for the control of the phase shifters can be implemented into the ceramic body in a compact arrangement. In contrast, organic materials offer a low permittivity down to values of $\varepsilon_r = 2$ in the GHz-range. In the area of the antenna elements, this material property is advantageous, as optimized radiation patterns at a high energy efficiency are generated. The organic foils having the antenna elements located on top are implemented into a cavity inserted in the first ceramic layer. Doing so, the length of the wire bonds is reduced to a minimum when ensuring low discrepancy in height in respect to the surface of the adjacent feed network.

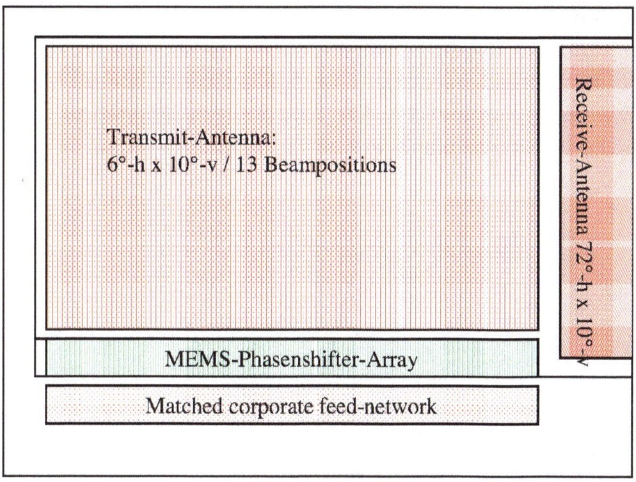

Fig. 1. Schematic of the RF front end with phased array antenna approach.

2.2 Phased Array Antenna with RF-MEMS Phase Shifters at 79 GHz

For the radar systems under consideration, namely automotive and wake-vortex detection radar, both receive and transmit antennas were first completely modelled, simulated and then fabricated. To finally assemble all sub-components of the electronically steerable antennas, several steps were performed in advance to characterize the individual parts of the antenna architecture (i.e. the fixed-beam antennas and the RF-MEMS phase shifters). The first step was the fabrication of fixed-beam antennas to determine their radiation patterns and to validate the excitation coefficients of the feed networks [3]. One example of these antennas is shown in Fig. 2, which illustrates the receive antenna for the wake-vortex detection radar.

The transmit antenna consists of 28 1D-antenna arrays fed in parallel by a corporate feed network. This arrangement results in a narrow, pencil-shaped beam characteristic, which can be steered in 21 discrete positions between ±30° in the final stage. The excitation coefficients used to feed the 2D arrays are determined by a so-called Remez-type algorithm. This algorithm computes the best approximation in the Tchebycheff sense of the ideal radiation characteristic and ensures the realization of the minimum side lobe level, while achieving a certain predefined beam width. The corporate feed network is fabricated with asymmetrical T-junction power dividers dimensioned to achieve the Tchebycheff aperture distribution. The feed network of the next generation of antennas will be realized on LTCC substrate, having a higher dielectric constant than the Rogers substrate currently used in order to minimize the radiation of the feed network. The location of the RF-MEMS phase shifters, which will be implemented in the next integration step to steer the antenna beam in the horizontal plane, are also indicated in Fig. 2.

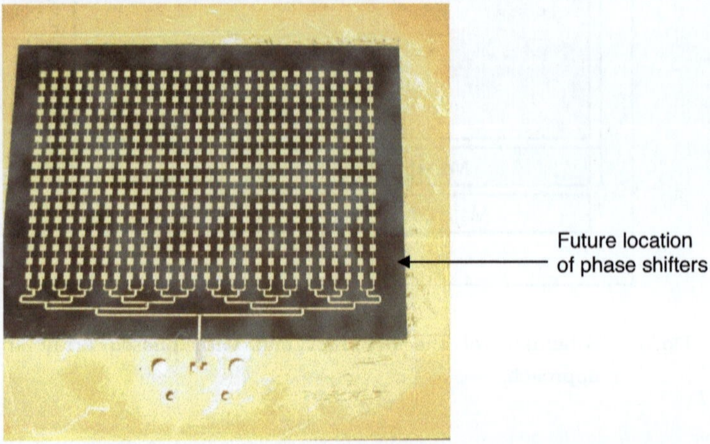

Fig. 2. Microstrip patch antenna array with beam widths of 3° in horizontal and 5° in vertical direction.

While designing and fabricating the fixed-beam antennas, effort was put in parallel in the realization of the RF-MEMS based phase shifters. To achieve the necessary performance in terms of beam-steering, 4-bit digital phase shifters were developed. These are fabricated in the EADS low-complexity RF-MEMS technology [4], which already demonstrated excellent microwave performance [5] with high reliability [6]. The RF-MEMS switches incorporated in the phase shifting circuits showed low insertion losses and high isolation at 77 GHz measured on wafer level. The complete phase shifters are already fabricated and are currently measured. A microscope photograph of a 180° bit in switched-line configuration is illustrated in Fig. 3.

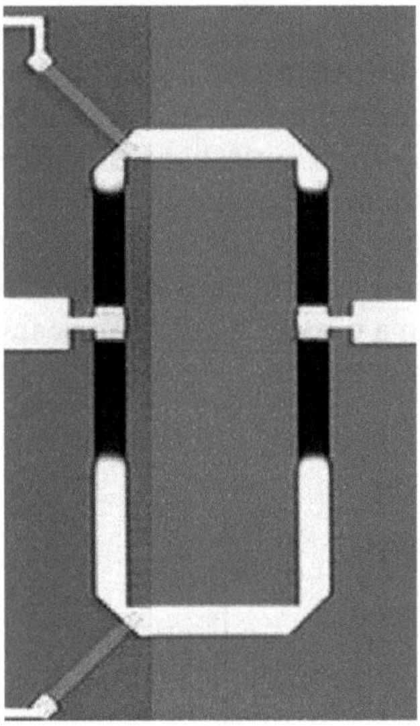

Fig. 3. Detailed view on the 180°-bit of a 4-bit RF-MEMS phase shifter.

3 Application and Product-Related Objectives

As mentioned above, the radar sensor is designed for surveillance purposes in the mid- and short range regime for automotive application scenarios. Applying marginal modifications to the design of the antennas and during operation, a sensor element with a similar design can be used for the detection of wake vortices. Another potential application scenario for the radar sensor is the surveillance of the operating area of robots, enabling a safe cooperation of individuals and robots in the same working place.

3.1 Automotive Radar for the Short- and Midrange Obstacle Detection

Basically, driver assistance systems need a narrow beam to detect objects at long distances and to reject reflections originating from objects located next to the road. To survey the traffic on several lanes in medium and short distances, however, a wide detection range is required. The movement of vehicles in the neighbourhood shall be monitored to support manoeuvres, such as lane change. Both tasks can be covered by a phased array antenna, which provides electronic beam steering of a narrow lobe over a wide detection range. A further potential field of application is the autonomous brake activation when collisions with pedestrians or other traffic participants are inevitable. For this purpose, the radar sensor scans in the narrow-beam configuration the traffic situation establishing in front of the car by the detection of the position and the relative velocities of the different obstacles in respect to each other. Besides further data originating from optical-based devices (e.g. cameras) these informations are supplied to an electronic control unit which activates under this condition the emergency brake modus. Fig. 4 shows the specified detection range of the radar sensor. Several lanes are monitored with 13 beams up to a distance of 40 m within a view of ±40°. The beams are activated sequentially. The angular resolution is 0.6° at an accuracy in depth of 15 cm.

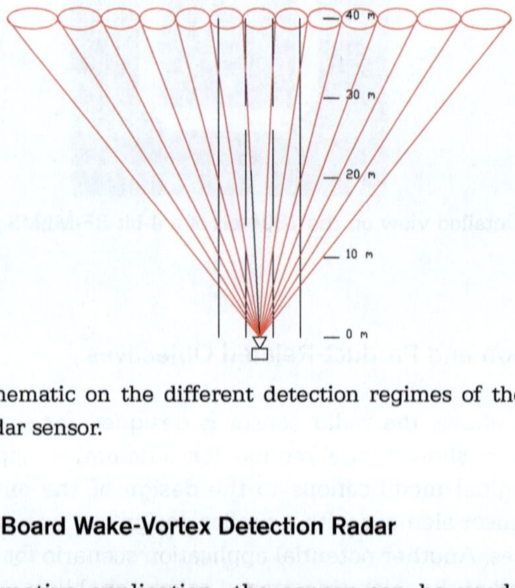

Fig. 4. Schematic on the different detection regimes of the automotive radar sensor.

3.2 Avionic On-Board Wake-Vortex Detection Radar

Beside the automotive application, the sensor can be used to detect wake-vortices of departing and landing aircrafts. This allows the optimization of the time intervals typically inserted between take-off and landing procedures. When implemented on-board, areas having a high vortex density are located and the aircraft is exposed to lower vibrational stress levels due to an adapted timing for e.g. take-off. Furthermore, additional information on the upcoming flight conditions is provided for the automatic aircraft stabilization. At bright weather conditions these hazardous turbulences in the air are detected with optical sensors based e.g. on LIDAR. Basically, the motion of the water contents is tracked independently of their physical condition, such as water droplets, snow or ice particles. The detection limit, however, is already reached at moderate raining (i.e. 1 mm/m^2 per hour) and snowing conditions. Fig. 5 shows the specified detection range of the corresponding radar sensor. In contrast to the automotive type, the view is slightly reduced to ±30°, but a higher number of beams (i.e. 21) with an aperture of 3° scan the front view. Furthermore, only a resolution in depth of 1 m is requested, as this information is of less importance. For this application, the velocity vector of each individual clutter cell is of interest, showing the shape and strength of the wake vortex. These informations on the actual state of the wake vortices is electronically evaluated and made visible to the pilot or to the decision makers working in the air safety department. In a more sophisticated scenario, the corresponding flaps at the aircraft are activated to perform a self-controlled stabilization during operation.

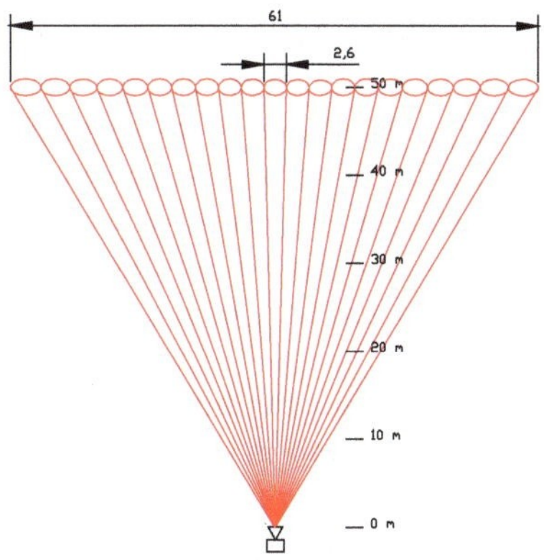

Fig. 5. Schematic on the different detection regimes of the airborne radar sensor.

4 System Design and Simulation Results

Independent of the application scenario, it is proposed to use a linear, frequency-modulated continuous wave (FMCW) radar with extremely high modulation frequency. The latter feature enables to measure the distance and the velocity of each pixel simultaneously. To determine the sign of each component of the velocity vectors and to increase the sensitivity a two-channel receiver is implemented (I/Q mixer) in the receiver electronics which provides information not only on the amplitude, but also on the phasing of the detected obstacles. Only the beam of the transmitting antenna is steered to keep the volume of the sensor module as low as possible. The receiving antenna covers with a broad beam the complete field of view. The angles which the objects take up in respect to the car are simply deduced from the position of the exciting beam.

4.1 Functional Block Diagram with GaAs- or SiGe Chip Set

In Fig. 6a, the functional block diagram of the radar sensor using electronic components based on GaAs is shown. First, the transmitting signal is generated by an oscillator at 12.6 GHz and subsequently linearly modulated in the frequency domain. The oscillator used offers an enhanced linearity as well as extreme low phase noise levels. The frequency output is stabilized using a phase-locked loop (PLL) approach, so that a control of the frequency is not possible when ultra-fast sweeps are applied. Under these conditions, the oscillator is free-oscillating. The oscillator output signal is transformed into the 76 to 77 GHz range and amplified. With the corporate feed network the signal is distributed across the phase shifters which control the antenna lines, thus realizing the beam steering. A portion of the transmitting signal is diverted by a directional coupler to provide an oscillating signal serving as reference for the receiver. The reflected signal is detected with the receiving antenna in a broad beam configuration covering the complete view. The use of a low-noise pre-amplifier increases the sensitivity of the complete system substantially.

After having passed the I/Q mixer the information on both the amplitude and the phasing of the reflecting obstacles is gained enabling a distinct determination of the velocity vectors which is not possible with a one-channel receiver. The mixer converts the signal directly into the base band. When applying the FMCW radar approach the frequency of the beat signals is proportional to the distance of the obstacles. Undesired reflections originating in the short range regime from a crosstalk of the antenna elements, the radiator grill or from the bumpers are eliminated from the signal path by a high pass filter after having passed the mixer. Next, the signal is amplified and filtered again with an element having a $1/f^3$-output characteristics compensating the free-space attenuation of the radar signal and hence, increasing the dynamic range of the radar sensor. This important feature is further amplified by the implementation of a software-controlled AGC (automatic gain control) amplifier which is mainly necessary for the automotive application scenario, as obstacles having a very strong reflection characteristics, such as loading door of trucks, as well as those with a low reflecting performance, such as individuals or plastic-based posts, may both occur in the view.

After amplification the signal is confined in its bandwidth by an anti-aliasing filter and loaded into an A/D converter. Finally, a FPGA (field programmable gate array) provides a pre-conditioning so that the information both on the distance and on the velocity is extracted for each single range gate from the raw data. The modulation and the beam steering applied during operation of the radar sensor are performed with a micro-controller which performs in addition the post-processing and delivers the object list to the driver assistance system. In Fig. 6b, the layout of a front-end is illustrated using components fabricated in low-cost silicon germanium (SiGe) technology. In contrast to the GaAs approach, the signal is directly generated in the millimetre wave band at 76 GHz. As the chips offer a differential output, the signals need to be transferred to the microstrip line with a balun. Besides the oscillator, the mixer elements are fabricated in SiGe technology. For the pre-amplifier, no prototype or commercial product is available. Therefore, this component is replaced by a GaAs chip.

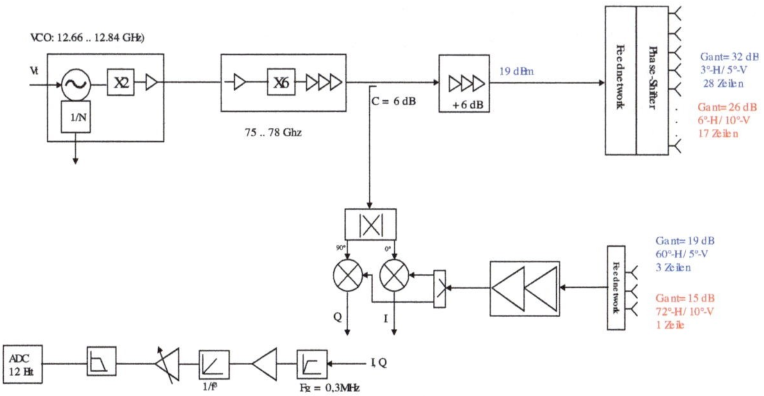

Fig. 6a. Functional block diagram when implementing GaAs components.

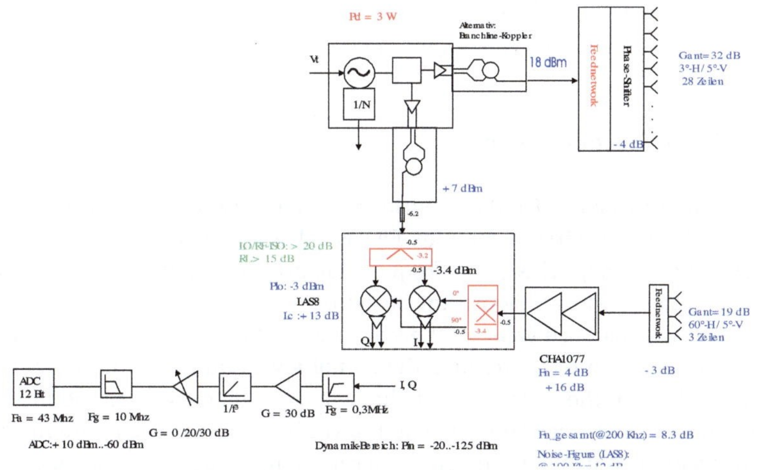

Fig. 6b. Functional block diagram when implementing SiGe components.

4.2 Signal Modulation and Timing

When applying the signal modulation a linear frequency modulation is applied with an ultra-fast slew rate. The corresponding time frame is typically in the 10 µs range. In Fig. 7, the modulation scheme is illustrated. The receiving signal is sampled and synchronized with the modulation frequency of the transmitter. The signal pre-condition is delayed by one modulation period. For an enhanced sensitivity of the system, 128 sweeps are averaged at each antenna beam. The evaluation of additional sweeps is taken into account to increase the resolution of the velocity vector which is derived from the change in phasing for each obstacle. The distance range covered by one beam is measured within a time frame of about 1 ms, so that the complete view in front of the automobile is scanned within a maximum time frame of about 13 ms.

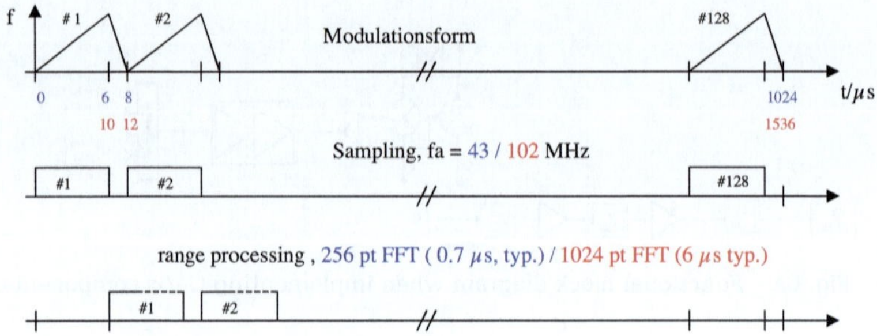

Fig. 7. Modulation scheme und timing for both application scenarios (blue: automotive, red: airborne application).

4.3 Signal Processing Concept

A two dimensional FFT (fast fourier transformation) is performed for signal pre-conditioning. First, a complex FFT is calculated from the data gained from each single sweep. Next, the frequencies are arranged in a matrix which is again subjected to an algorithm based on FFT. This final matrix provides in the rows the distance values, whereas in the columns the velocity is displayed. Furthermore, the FFT averages the signal levels due to an integration procedure. Doing so, the sensitivity of the system is increased, as the influence of uncorrelated noise is reduced. The matrix built up for each antenna beam is either suited to display the corresponding amplitudes or to generate a velocity-related graph.

4.4 Simulation of the Received Signals Gained from Predefined Objects

To estimate from pre-defined objects and under different weather conditions the quality of the back-scattered signals serving as input for the radar sensor, extensive simulations were performed. The receive signal is conditioned according to the above mentioned procedure. Besides the characteristic values associated with the electronic components, shown in the functional block diagrams (see Figs. 6a and 6b), critical parameters of real devices, such as the non-linear output behaviour of the oscillators, damping factors, antenna parameters, noise figures of the receiver, are considered in the calculations. Fig. 8 shows a typical result. In this simulation, three objects are assumed having the following parameters: Object 1 is located at a distance $d_1 = 10$ m with a velocity $v_1 = 40$ m · s^{-1}. Object 2 (3) has values $d_2 = 15$ m ($d_3 = 40$ m) and $v_2 = 4$ m · s^{-1} ($v_3 = 50$ m · s^{-1}), respectively. In all cases, the minimum reflectance of a sphere with a diameter of 10 cm is assumed. In Fig. 8, the x-axis shows the velocity, the y-axis the distance and the z-axis the signal height. At the origin of the coordinate system, the reflection of the antenna is observed which is eliminated by the high-pass filter later on. Basically, all three objects can be well detected due to an outstanding signal-to-noise ratio.

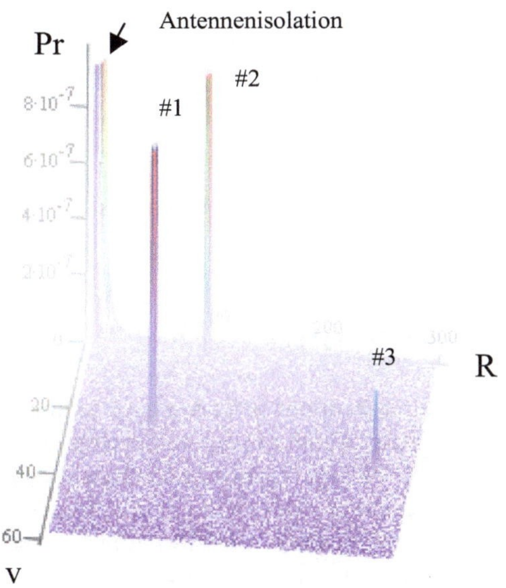

Fig. 8. Simulation results on the detection performance of the three objects.

4.5 Functional Verification of Signal Processing Concept and System Performance Using an Experimental Radar

To verify the concept for signal conditioning and to validate the tracking of the hydrometeors via their velocity profiles, a field test was performed applying an experimental radar. For this basic study, a radar module whose antennas still need to be steered mechanically, thus having a low degree of integration, was fixed to the bumper. The presence of hydrometeors was simulated by a water jet. Additional objects with known reflectance were placed into the view of the radar sensor. Fig. 9 shows the set-up used for testing and a typical measurement result where the different colours are attributed to varying velocities of the objects: objects in rest are displayed in green and the water jet containing moving water droplets is in blue. The information on the amplitude is not visible in this configuration.

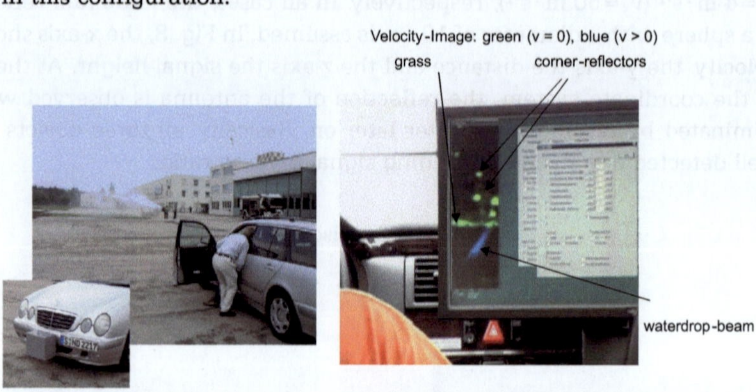

Fig. 9. Field-test with a state-of-the art radar sensor for automobiles.

5 Conclusions and Outlook

In this study, basic considerations are presented to develop an advanced radar sensor with phased array antennas for automotive and airborne applications targeted to operate at 79 GHz. Using an electronic beam steering approach, it is feasible to implement a radar module which does not need to be moved mechanically, thus giving space in the automobile for additional environment sensor systems, such as a camera. For an overall compact design, LTCC technology is combined with organic materials to have on the one hand the possibility for a high degree of integration and on the other hand to modify dielectric properties to a maximum extend. First results on core components, such as the antenna performance and the MEMS–based switching elements, are very promising and give confidence that prototypes can be successfully fabricated.

To design the read-out and the control electronics functional block diagrams were built up for both potential scenarios using either GaAs or SiGe chip sets. Furthermore, the receiving signals were simulated applying realistic values for the implemented components resulting in an acceptable signal-to-noise ratio. A field test using a mechanically steered automotive radar module at 76 GHz proves the potential to detect hydrometeors. Therefore, it can be concluded that radar sensors in this frequency range are also well suited for airborne applications.

Acknowledgements

This work was performed within the RADARAUGE project (http://www.radarauge-project.com/) financially supported by the German Federal Ministry of Education and Research (BMBF). This support is gratefully acknowledged.

References

[1] A. Bittner, H. Seidel, T. Haas, and U. Schmid, "Local modification of fired LTCC substrates for high frequency applications", accepted at the 12th AMAA conference, Berlin, 2008.

[2] S. Klein, V. Ziegler, U. Prechtel, A. Würtz, H. Seidel, U. Schmid, "Tungsten-Titanium as advanced material for RF-MEMS switches", accepted at the 12th AMAA conference, Berlin, 2008.

[3] W. Gautier, A. Stehle, B. Schönlinner, V. Ziegler, U. Prechtel, and W. Menzel, "Antenna arrays for RF-MEMS based 77 GHz on-board wake vortex detection sensor", 2nd Int. ITG Conference on Antennas, Munich, Germany, 28.-30. March 2007.

[4] C. Siegel, V. Ziegler, B. Schönlinner, U. Prechtel, and H. Schumacher, „Simplified RF-MEMS switches using implanted conductors and thermal oxide", European Microwave Week, European Microwave Conference, Manchester, UK, 10.-15. September 2006.

[5] C. Siegel, V. Ziegler, U. Prechtel, B. Schönlinner, and H. Schumacher, „A Ka-Band RF-MEMS phase shifter approach based on a novel dual-state microstrip line", 10th European Microwave Week, European Microwave Conference, pp. 1221-1224, Munich, Germany, 08.-12. October 2007.

[6] A. Stehle, C. Siegel, V. Ziegler, B. Schönlinner, U. Prechtel, S. Thilmont, H. Seidel, U. Schmid, "Low complexity RF-MEMS switch optimized for operation up to 120°C", 10th European Microwave Week, European Microwave Conference, pp. 1229-1232, Munich, Germany, 08.-12. October 2007.

Richard Körber
Astyx GmbH
Lise Meitner Str. 2a
85521 Ottobrunn
Germany.
r.koerber@astyx.de

Volker Ziegler
EADS Innovation Works
Dept. IW-SI Sensors, Electronics & Systems
81663 München
Germany
volker.ziegler@eads.net

Ulrich Schmid
Saarland University
University Campus 15.1
66123 Saarbrücken
Germany
u.schmid@lmm.uni-saarland.de

Keywords: imaging radar, phased array, RF-MEMS phase shifter, LTCC, organic substrates.

Tungsten-Titanium as Advanced Material for RF-MEMS Switches

S. Klein, H. Seidel, U. Schmid, Saarland University
V. Ziegler, U. Prechtel, EADS Innovation Works
A. Würtz, Atmel Germany GmbH

Abstract

In this paper we present first results on the development of high temperature stable tungsten-titanium (WTi) based MEMS cantilevers, using conventional sputter deposition technique and wet chemical etching. By synthesizing two WTi layers with slightly different, but tailored film properties an out of plane deflection of the cantilevers is achieved. This core component is most favourably used in RF-MEMS switches actuated electrostatically. Based on this design, a high isolation in the up-state and a low insertion loss in down-state are guaranteed. Furthermore, the temperature stability of WTi thin films is determined and evaluated up to a maximum annealing temperature of 450 °C, appearing during hardware realisation. The results gained so far give confidence that this type of MEMS cantilevers is well suited for the proposed application scenario.

1 Introduction

In recent years, one of the main trends to improve car safety is to equip automobiles with radar-based on-board driver assistance systems. In order to exploit their full potential two basic requirements are essential: a narrow beam is requested for long distances and a broader detection range for short or medium distances. These application-oriented requests can be met with a radar beam, which is electronically steered. For an adequate control of the different antenna elements with respect to their phasing, the use of RF-MEMS switches represents an advanced approach for the realisation of phase shifters with optimized performance due to a low insertion loss in down-state and a high isolation in up-state of the micromachined component [1]. Since sticking is one of the most frequent reasons for the failure of movable MEMS devices, especially when fabricated in surface micromachining, a hermetic housing is preferentially targeted to protect these structures from humidity and hence, to achieve a good long term stability. Most of the commonly on-wafer level housing techniques, however, providing a hermetic packaging, require temperatures starting well above 250°C. As demonstrated in a previously published study on the temperature stability of aluminium-based cantilevers, reliable operation up to 120°C can be achieved with this material combination [2]. In addition, packaging temperatures up to 250°C can be withstood using the current alloy composition. But, with the objective to monolithically integrate the RF-MEMS switches with silicon germanium (SiGe) devices and to use high-temperature hermetic packaging processes, a material with higher temperature stability is needed to conserve the intrinsic film stress implemented intentionally during deposition. Due to its high melting point tungsten (W) is regarded to be well suited for this purpose. Since the adhesion of tungsten on silicon dioxide is a critical parameter, an alloy made of tungsten and titanium (Ti) is used, accepting somehow lower values for the electrical conductivity and the thermal stability compared to pure tungsten.

It is the objective of this study to present first results on the development of high temperature stable, WTi-based, cantilevers. The technology used for fabrication is reported. Furthermore, the temperature stability of sputter-deposited WTi thin film is determined and evaluated up to a maximum annealing temperature of 450°C.

2 Fabrication Process

The devices are fabricated applying a surface micromachining approach on thermally oxidised silicon wafers. First, a thin layer of aluminium (Al) is deposited by DC magnetron sputtering on the front side of the wafer. This layer is used as the sacrificial layer during device fabrication and hence, its height strongly influences the actuation voltage required for switching. Next, the oxide on the backside is removed with buffered hydrofluoric acid. After that the sacrificial layer is patterned with standard photo lithography and a subsequent wet etch in phosphoric acid. The functional layer consists of two WTi layers each 1 µm thick, which are sputter deposited from a tungsten (90%)-titanium (10%) target. Defined variations in deposition parameters cause the layers to be slightly different in nature. Thus, the biaxial film stress is influenced, resulting in the desired deflection after the release of the structures. Controlling the sign and the level of the film stress in each layer is most challenging during the fabrication process for the reproducible realisation of cantilevers, as a predefined bending behaviour is of utmost importance for the targeted RF-MEMS switches. Again, the patterning of the WTi layers with a thickness of 2 µm in total is done with standard photo lithography and a wet etch process. A hard mask is required for this procedure, since the etching is done in hydrogen peroxide. For the final release, the sacrificial layer is exposed to phosphoric acid. In the same processing step the hard mask is removed, as the sacrificial layer as well as the hard mask consists of the same material. After etching, the wafers are first purged in deionised water and then swilled in hot propanol to reduce in-process sticking [3]. Due to the high spring force of the beams a more sophisticated release process, such as offered by a critical point dryer, is not absolutely necessary.

3 Results

3.1 One-Sided Clamped Cantilevers

Figs. 1 and 2 show an example for a released beam structure. Fig. 1 is taken with a scanning electron microscope (SEM) at a magnification of 350 and Fig. 2 shows a scan done with a white light interferometer (WLI). As desired, the cantilevers bend away from the surface due to the difference in film stress of both layers. So the air gap between substrate and cantilever increases from the anchor to the free end and this leads to the high isolation in up state. For evaluation purposes cantilevers with two different widths w, 150 µm and 200 µm, and different lengths L ranging from 100 µm up to 600 µm with a step size of 50 µm have been realised. The corresponding tip deflections cover a spectrum from 4 µm (i.e. at L = 100 µm) up to 80 µm at 550 µm long cantilevers. For an enhanced performance during release and an easy access for inspection the sacrificial layer was chosen to 400 nm in the first run.

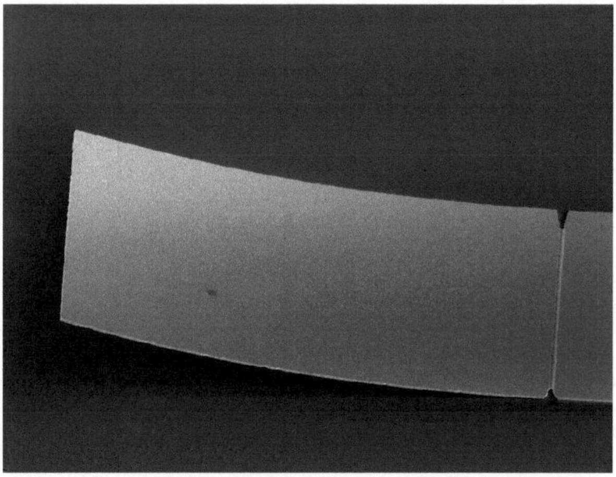

Fig. 1. SEM picture of the released tungsten-titanium cantilever on SiO_2.

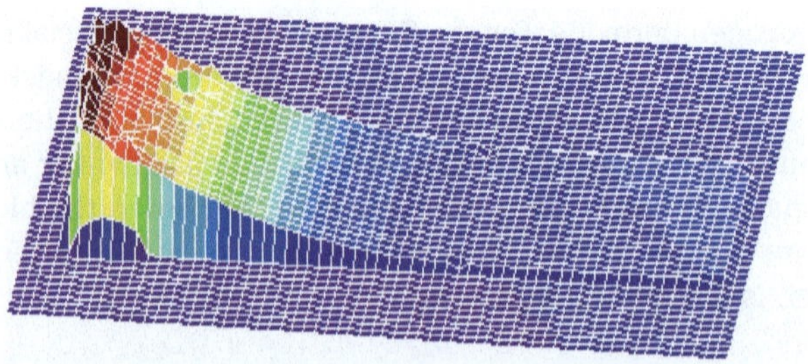

Fig. 2. Surface profile of the cantilever presented in Fig. 1.

The length and the width of the cantilever shown in Figs. 1 and 2 are 350 µm and 150 µm, respectively. The maximum tip deflection determined with both methods is about 54,5 µm. A typical plot of the curvature extracted from the measurement with the WLI can be seen in Fig. 3. Typically, the deflection has a circular shape. This cannot be seen in the plot based on the different scaling of the axes. Due to the low aspect ratio when comparing the thickness of the sacrificial with the functional layer, the transition region between the cantilever and the anchor is clearly to observe (see Fig. 4). Besides this influence, the WTi looks fragile in this area, as the film stress in the active layer is in addition very high (see Chap. 3.2). Applying an optimized design with a reduced gap height in the order of 100 nm and below while keeping the WTi thickness as fixed, this weak point can be eliminated.

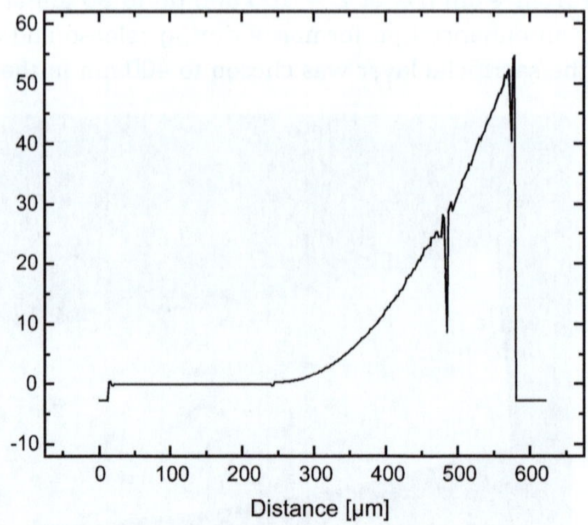

Fig. 3. Curvature along the centerline of the beam.

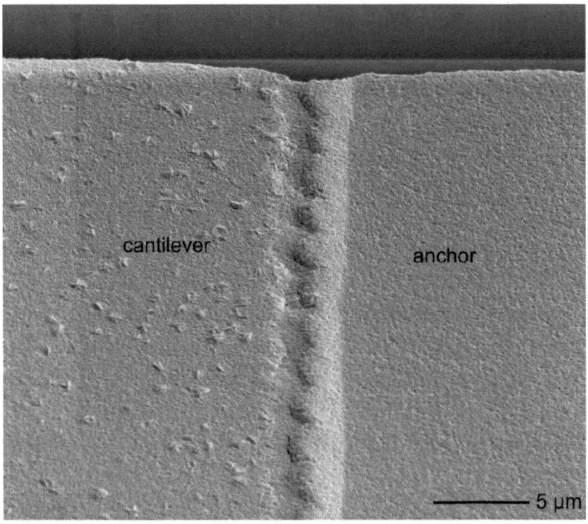

Fig. 4 SEM micrograph of the transition region between released structure and anchor.

Despite this large air gap the cantilevers can be electrostatically activated resulting in actuation voltages of 60 V and above. According to [4] the pull-in voltage of cantilevers with residual stress gradients is described by

$$V_{pull-in} = 5.51 \times 10^{-4} \sqrt{\frac{E_{eff} t^3 \left(h + \delta_0(\Gamma)\right)^2}{\varepsilon_0 L^3 \left(L + 3(h+t)\right)}}, \tag{1}$$

where E_{eff} denotes the effective Young's modulus, L and t the length and the thickness of the cantilever, ε_0 the electrical permittivity in vacuum and $\delta_0(\Gamma)$ the tip deflection. When reducing only the gap size h from 400 nm to 100 nm while leaving all other parameters unchanged, the pull-in voltage is assumed to decrease by a factor of approximately two. Thus, actuation voltages in the order of 30 V seem to be feasible. For each consecutive switching operation the sign of the voltage has to be changed. Otherwise, the beam structure is not pulled downwards reliably because of charging effects in the silicon oxide layer located below (see Ref. [2] and references cited herein).

3.2 Temperature Stability

As mentioned in the previous chapters RF-MEMS switches are very sensitive to environmental influences, such as humidity and contaminants. Therefore, a hermetic on-wafer level packaging is the preferred choice for sealing these devices. Since all conventional hermetic packaging techniques (e.g. based on glass frit) require temperatures in the range of 400°C, the tungsten-titanium thin films ideally have to be resistant to any impacts on material properties arising from temperature. At least, the change in film properties needs to be predictable and reproducible. Consequently, the parameter influencing the device performance predominantly as a function of annealing temperature is the film stress and hence, needs to be investigated in more detail.

For these analyses a set of different WTi thin films have been prepared by sputter deposition on double-side polished 4" silicon wafers. The synthetization parameters have been varied to a maximum extend to evaluate the performance of thin films with different properties in respect to their temperature stability. The values applied for thin film preparation are shown in the Tab. 1.

	Deposition Parameters			
Power [W]	*100*	*250*	*500*	*1000*
Pressure[mbar]	*3 10^{-3}*	*5 10^{-3}*	*7 10^{-3}*	*9 10^{-3}*
Ar-Flow [sccm]	*30*	*50*	*70*	*90*

Tab. 1. Parameters for WTi thin film deposition.

To cover the complete range of possibilities when combining the three most important sputtering parameters among each other, a bundle of 64 wafers would be needed. To save time and money, but keeping the loss of information on the interaction of the different parameters in respect to their influence on film growth and film stress as low as possible, 16 wafers have been prepared using the approach proposed by the Taguchi method. The thickness of each thin film is fixed to 1 µm. Applying the well-established "wafer bow" measurement technique to determined the corresponding curvature, the biaxial film stress is gained using Stoney's formula [5]. After measuring the stress in the "as-deposited" condition all wafers are put into a muffle kiln for annealing and analysed subsequently to any change in film stress after having cooled down to room temperature. Following this sequence, the samples are heated up to 450°C with a step size of 25°C starting at 50°C. The dwell time at each peak temperature is 30 minutes.

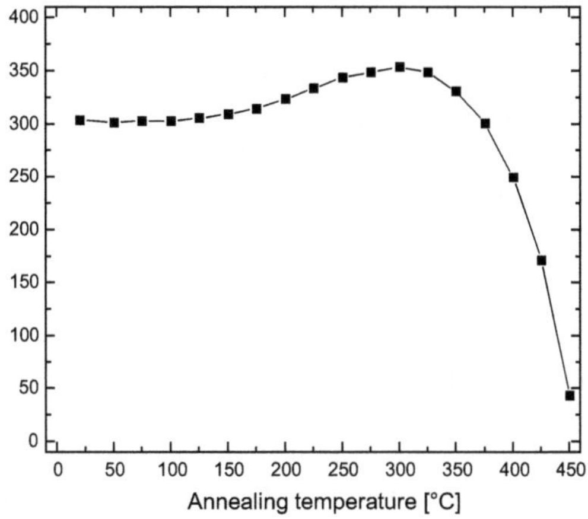

Fig. 5 WTi thin film with tensile stress.

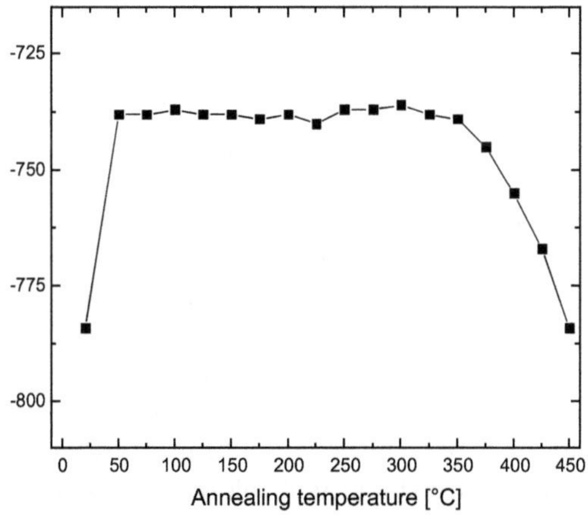

Fig. 6 WTi thin film with compressive stress.

Depending on the deposition parameters tungsten-titanium thin films with either tensile or compressive stress can be realised (see Figs. 5 and 6). Besides the difference in sign, the temperature behaviour of the film stress varies to a great extent when applying different sputter parameters. Most of the samples prepared show an annealing behaviour whose characteristics are similar to that shown in Fig. 5. Starting from room temperature the stress is either more or less constant up to a temperature of about 350°C before reaching a local maximum in the upper annealing range. Increasing the annealing temperatures even further, the film stress starts to decrease for all samples investigated, independent whether they show originally tensile or compressive stress behaviour. For samples with a low tensile stress in the "as-deposited" state it is even possible that they change from tensile to compressive stress values. Compared to the aluminium based cantilevers, the WTi films exceed their temperature limitation substantially and make possible the realization of cantilevers with a film stress that is almost constant over the targeted temperature range.

5 Conclusions and Outlook

In this study cantilevers based on tungsten-titanium were presented. Depending on the geometrical dimensions tip deflections up to 80 µm were realized at a film thickness of 2 µm. With voltages alternating in their polarity and applied across the substrate the beam structures could be activated. The absolute values of the switching voltages were larger than 60 V due to an air gap of 400 nm. According to theoretical calculations, however, these values can be reduced by a factor of two when decreasing the gap height to 100 nm.

Besides the fabrication of the cantilever itself, the temperature behaviour of WTi thin films was evaluated up to annealing levels of 450°C. It was demonstrated that in the "as deposited" state the residual film stress can be tailored in a broad range using different deposition parameters. Doing so, thin films with either tensile or compressive film stress can be realised. Basically, the samples can be divided into two types: the first show a temperature behaviour which is relatively constant over the whole temperature range. In contrast, the second type has an intrinsic stress that is constant up to 200°C before increasing when exposed to 350°C. Above, the film stress decreases at annealing temperatures up to 450°C. In the near future further investigations have to focus on the temperature stability of released cantilevers up to temperatures of 450°C and on the high frequency characterisation of the corresponding switches. Based on the results gained so far, however, it seems to be feasible to realize one-sided clamped high temperature stable cantilevers representing the core component of micromachined RF switches.

Acknowledgements

This work was performed within the RADARAUGE project (http://www.radarauge-project.com/) financially supported by the Federal Ministry of Education and Research (BMBF) under contract number 16SV2080. This support is gratefully acknowledged.

References

[1] C. Siegel, V. Ziegler, U. Prechtel, B. Schönlinner, H. Schumacher, "A Ka-Band RF-MEMS Phase Shifter Approach Based on a Novel Dual-State Microstrip Line", 10th European Microwave Week, European Microwave Conference, pp: 1221-1224, Munich, Germany, 08.-12. October 2007.

[2] A. Stehle, C. Siegel, V. Ziegler, B. Schönlinner, U. Prechtel, S. Thilmont, H. Seidel and U. Schmid, "Low complexity RF-MEMS switch optimized for operation up to 120°C", Proc. 2nd European Microwave Integrated Circuits Conference, pp. 431-434, 8.10.-10.10, Munich, 2007.

[3] C. Siegel, V. Ziegler, U. Prechtel, and H. Schumacher, „Low-complexity RF-MEMS technology for microwave phase shifting applications", German Microwave Conf., pp. 13-16, Ulm, Germany, 5.-7. April 2005.

[4] M. Lishchynska, N. Cordero, O. Slattery, and C. O'Mahony, "Modelling electrostatic behaviour of microcantilevers incorporating residual stress gradient and non-ideal anchors", J. Micromech. Microeng. 15 (2005) S10-S14.

[5] G. G. Stoney, "The tension of metallic films deposited by electrolysis," Proc. R. Soc. London Ser. A 82 (1909) 172-175.

Stefan Klein, Helmut Seidel, Ulrich Schmidt
Saarland University
Chair of Micromechanics, Microfluidics/Microactuators
University Campus, Building A5.1, Room 1.09
66123 Saarbrücken
Germany
s.klein@lmm.uni-saarland.de
seidel@lmm.uni-saarland.de
u.schmidt@lmm.uni-saarland.de

Volker Ziegler, Ulrich Prechtel
Eads Innovation Works, Dept. IW-SI
81663 München
volker.ziegler@eads.net
ulrich.prechtel@eads.net

Alida Würtz
Atmel Germany GmbH
Theresienstr. 2
74072 Heilbronn
Germany
alida.wuertz@hno.atmel.com

Keywords: tungsten-titanium, stress, curvature, RF-MEMS, annealing behaviour

Local Modification of Fired LTCC Substrates for High Frequency Applications

A. Bittner, H. Seidel, U. Schmid, Saarland University
T. Haas, Micro Systems Engineering GmbH

Abstract

In this study, a novel process is presented to generate a defined and homogeneous degree of porosification in fired LTCC (low temperature co-fired ceramics) substrates. For this purpose, a phosphoric-based acid is used which represents a standard wet chemical etchant in MEMS and microelectronic industry. This surface-near process is very attractive for the realization of selected areas on conventional LTCC substrates having modified dielectric properties, especially for high frequency applications. From morphological investigations, this portion of the glass matrix enveloping the Al_2O_3 particles is very important to enable the penetration of the acid into the LTCC body. Increasing the time of the etch attack at a given bath temperature, pores are preferentially formed due to a selective removal of the Al_2O_3 particles compared to the residual glass matrix. High frequency measurements of the permittivity using ring oscillators showed a significant reduction compared to the original bulk value.

1 Introduction

In recent years, LTCC (low temperature co-fired ceramics) has attracted much attention as technology platform for the realization of advanced hardware solutions both on device and substrate level. This is in particular due to the multilayer approach, the possibility to implement vias with a low sheet resistance based on Au, Ag or Ag/Pd and the integration of passive electronic components (i.e. inductors, resistors and capacities) into the ceramic body, thus making use of the 3rd dimension [1,2]. The latter advantage enables the arrangement of electronic components in a compact way within a gas-proof body. Therefore, they are well protected from environmental impacts when operated under harsh environmental conditions, such as high temperatures [3]. Basically, LTCC is a glass-ceramic consisting of a glass matrix in which aluminium oxide particles with a typical diameter in the range of 2-3 µm are embedded as a filler material. For metallization purposes, thick film printing technique is the standard technology used. Due to the low sintering temperatures with peak levels in the range of 850°C the complete assembly of filled vias and printed structures are fired in one single step with the substrate which is in this status of the fabrication flexible (i.e green sheet). By the liquid sintering process the soft sheets are densified to form the ceramic body monolithic in nature.

LTCC is most favourably used as substrate for micromachined devices and systems operated at high frequencies typically ranging up to the microwave region. Although there are also other types of high-density, multilayer substrates available, based on organic laminates, further outstanding features of the LTCC for this field of application are the excellent thermal conductivity compared to organic materials and a coefficient of thermal expansion close to silicon [4]. Compared to a standard organic substrates reinforced with a glass-fibre based component, such as FR-4 or RT/Duroid, the dielectric losses are low. In contrast, the relatively high permittivity of $\varepsilon_r = 7 - 8$ is disadvantageous for some application scenarios, such as the performance of microwave antennas directly arranged on the LTCC surface. To avoid this drawback either a combination of LTCC with a local application of a low-*k* organic material [5] or the modification of the LTCC substrate itself is targeted. Beside the modification of the glass-matrix and the crystallization behaviour [6, 7], the generation of a defined porosity is the most commonly used approach to reduce the dielectric constant and the dielectric losses of materials [8].

It is the objective of the paper to report about a novel process to generate locally a defined porosity in LTCC being in the fired state. Up to now, a maximum penetration depth for the porosification process of about 40 µm below the substrate surface is demonstrated. Phosphoric-based acid is used which is a well-established chemical product used for the patterning of aluminium-based strip lines within the fabrication process of micromachined devices. The process and hence, the degree of porosification and the corresponding penetration depth, can be controlled very easily by monitoring the etch time and the temperature of the etchant at a given bath concentration. The impact of these relevant parameters on the modified LTCC microstructure is investigated via SEM (scanning electron microscopy) and FIB (focused ion beam) technique. A successful application of thick-film based test patterns on the modified LTCC is demonstrated. High frequency measurements in the GHz-range using ring resonators indicate a substantial decrease in permittivity.

2 Experimental Details

To study the porosifiaction process, commercially available LTCC substrates (DP 951 AX) from DuPont was used. The blank sheets were laminated at a pressure of 20 GPa and fired at a peak temperature of 850°C for 30 min. in batch furnace. Further details of the fabrication process can be found elsewhere [9]. After co-firing, a compound material was generated consisting of a glass matrix with different crystalline and chemical phases in which Al_2O_3 particles with a typical size in the µm-range are implemented (see Fig. 1). Basically, the DuPont 951 LTCC tape consists of corundum, anorthite and a glass phase consisting of a lead silicate glass after firing [10].

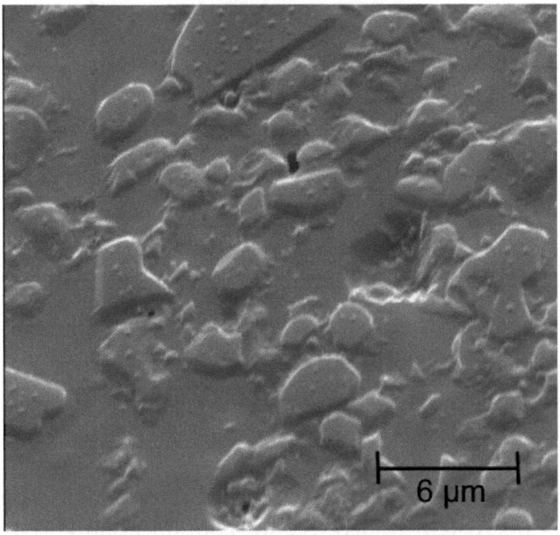

Fig. 1. SEM top view on a conventional LTCC (DP 951) substrate. To highlight the microstructure, the LTCC substrate is dry etched for several minutes applying an argon flux, thus removing the glass coverage from the surface.

To chemically attack the LTCC substrates, phosphoric acid with a purity of 85 vol% is used. The bath temperature is closed-loop controlled on a hotplate purchased from Heidolph Instruments and varied between 90 and 130°C. Dwell times at peak temperature were between 1 to 8 h. Finally, the samples were rinsed in deionized water and dried with purged air.

To measure the penetration depth of the porosification process, the specimens were embedded into a resin matrix (Demotec 30), cross grinded on a standard turntable from Struers and analysed via SEM technique (LEO 435VP). To investigate the micro gaps and pores generated after wet chemical etching in more detail, a SEM in a Dual Beam FIB - FEI Strata DB235 workstation was applied. In these cases, cross-sectioning was done by means of FIB technique. The surface roughness was measured with a profilometer (Alphastep 200) from Tencor Instr. At a fixed scan length of 2 mm, the values for the maximum roughness depth R_{max} and the average roughness R_a are determined.

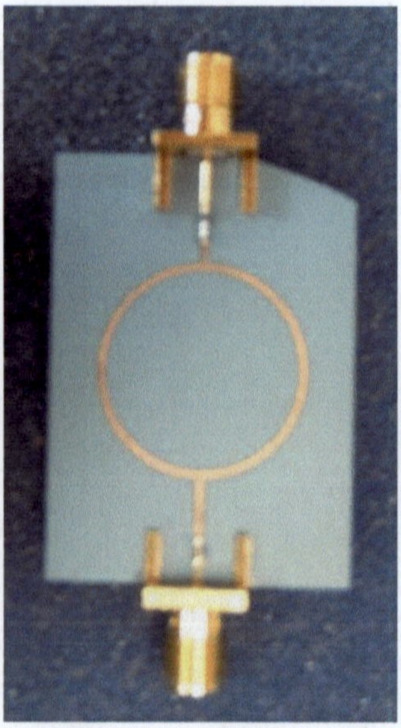

Fig. 2. Typical sit-in for the high frequency characterization of the substrates.

As shown in Fig. 2, the permittivity of the different LTCC substrates was measured using a microstrip ring resonator configuration [11]. Sputter deposition technique and standard lithography followed by a wet chemical etching process were applied for the realization of the copper-based test structure. The gap between the ring resonator and the feed lines is about 200 µm [see Fig. 2]. The electromagnetic wave is capacitively coupled into the resonant structure and extracted via a second feed line located on the opposite side. Standard SMA connectors are soldered carefully to the thin film metallization. The resonance spectra of the S-parameter S_{21} are measured with an Agilent E8363B network analyser in the range between 1 to 10 GHz. For an ideal loss free resonator the resonance frequency f_0 is an integer multiple of half of the ring perimeter p being in this case

$$f_0 = \frac{2c}{np} \tag{1}$$

Due to the dielectric properties of the substrate material the resonance peaks are shifted so that the effective dielectric constant can be calculated according to

$$\varepsilon_{r,eff} = \left(\frac{n \cdot c}{2\pi \cdot r_m \cdot f_{res}} \right)^2 \tag{2}$$

where the quantity n is an integer, c denotes the speed of light in vacuum, r_m the radius and f_{res} the measured resonance frequency. The relative dielectric constant of the LTCC material is derived from the effective dielectric constant and the effective width of the microstrip line:

$$\varepsilon_r = \frac{2 \cdot \varepsilon_{r,eff} + M - 1}{M + 1} \quad \text{with} \quad M := \left(1 + \frac{12 \cdot h}{W_{eff}} \right)^{-0.5} \tag{3}$$

where h is substrate thickness and W_{eff} the effective microstrip width, respectively. The latter quantity is determined according to

$$W_{eff} = W + \frac{t}{\pi}[1 + \ln(\frac{2 \cdot h}{t})] \tag{4}$$

where t denotes the conductor thickness and W the microstrip width [11].

3 Results and Discussion

In Fig. 3, the porosification depth d_p in fired LTCC substrates is known as a function of time t at different bath temperature levels T_b of the phosphoric acid is shown. A minimum value for T_b of 90°C is required to obtain in a time frame of about 1 h a detectable porosification depth of about 300 nm. As expected, d_p increases when enlarging the duration for the etch attack at a given bath temperature. Increasing T_b has a similar impact on d_p while keeping the parameter t as fixed. At the onset of the etch attack, d_p and hence, the porosification rate, is relatively low at T_b= 90 and 110°C. In the range of 2 to 5 h, the corresponding values of d_p show a linear relationship with t before reaching a bath-temperature dependant saturation level. Excluding under these conditions a period of about 2 h at the beginning to activate the porosification process, the curves obey in principle a functional characteristics, well-known e.g. for the different growth regimes of silicon dioxide when subjected to a standard wet oxidation process of silicon substrates [12]. This demonstrates that in the linear regime the etch attack for the porosification process of LTCC is reaction-limited, while in the saturation regime, it is diffusion-controlled. Increasing the bath temperature to 110°C, the porosification process is activated above average resulting in an almost linear relationship between d_p and t from the very beginning of the process. In comparison to the lower bath temperatures investigated, a saturation regime establishes after a period of about 5 h resulting in a maximum porosification depth of about 40 µm. From the application-oriented point of view, this wet etching process provides a sufficient depth promising a substantial reduction of dielectric properties, such as the permittivity, for high frequency applications. In Fig. 4, a typical cross-sectional view on the surface-near porosification is shown used to extract the data presented in Fig. 3. Besides the determination of d_p, a constant value for this important porosification parameter is demonstrated on technical-relevant length scales, providing a homogeneous reduction in dielectric properties across a given area.

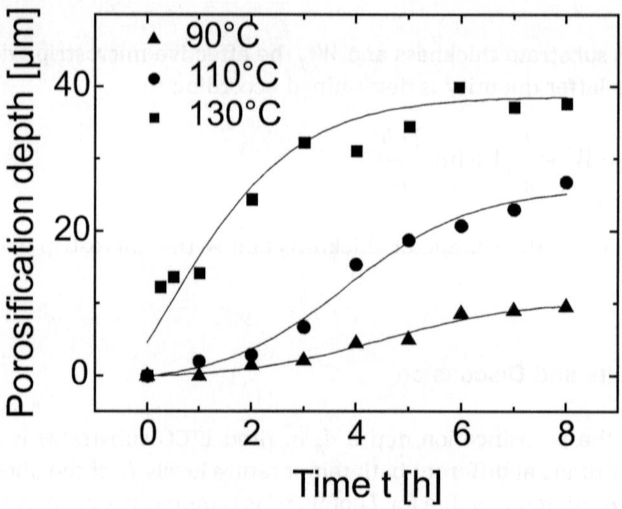

Fig. 3. Porosification depth d_p in LTCC (DP 951) as a function of time t at different bath temperatures of the phosphoric acid.

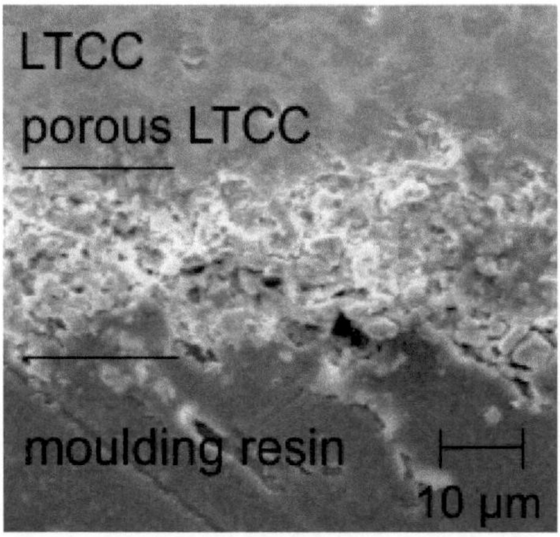

Fig. 4. Typical result when using a phosphoric acid at 110°C for 5 h to generate a porous microstructure in LTCC.

To investigate the nature of the porosification process and the corresponding microstructures in more detail, FIB technique in combination with SEM analyses were applied. For comparison purposes, Fig. 5a shows the topography of an "as manufactured" DP 951 LTCC substrate. The surface is highly undulated in the µm-range (i.e. $R_{max}\sim$ 1.7 µm and $R_a\sim$ 0.4 µm) due to the implemented Al_2O_3 particles and covered with the glass matrix based on the liquid sintering process. After exposure to the phosphoric acid for 3 h at 90°C, the glass matrix is locally attacked on the surface forming gaps and pores with dimensions well below 1 µm (see Fig. 5b). Increasing t to 8 h, the glassy top coverage is almost completely removed giving view on the Al_2O3 particles located below (see Fig. 5c). After etching 8 h at 130°C, the glass phase totally disappears (see Fig. 5d). Furthermore, the mean gap size is substantially increased indicating the penetration of the wet chemical into the body of the LTCC. It is worth mentioning that despite the removal of the top layer the Al_2O_3 particles still adhere sufficiently to the substrate due to only a partial removal of the glass matrix enveloping each single grain.

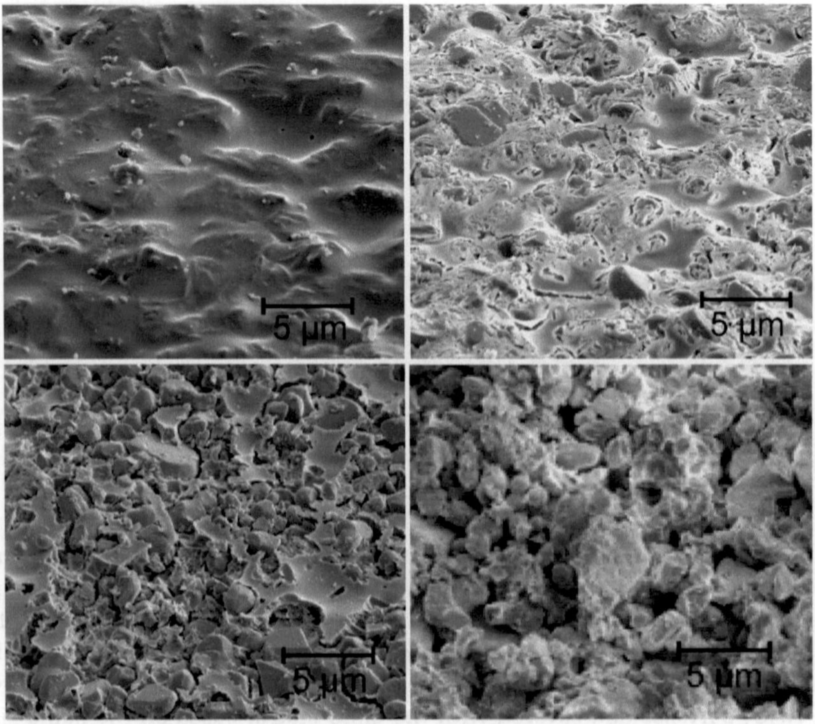

Fig. 5. Surface morphology of LTCC in different modifications:
a) "as manufactured" (upper left),
b) after etching for 3 h at T_b= 90°C (upper right),
c) after etching for 8 h at T_b= 90°C (lower left),
d) after etching for 8 h at T_b=130°C (lower right).

In Fig. 6a, a cross-sectional view on the porous layer is given after an etch attack at T_b= 90°C for 8 h. The assumption that the phosphoric acid penetrates via gap-sized structures into the body of the LTCC substrate is confirmed. This is especially obvious at the proceeding porosification front in the LTCC body where this portion of the glass matrix surrounding the Al_2O_3 particles is preferentially etched. Close to the surface, the formation of larger pores is indicated, as in this long etching regime, the Al_2O_3 grains are now selectively etched compared to the residual glass matrix. When increasing T_b, but decreasing t to 2 h, only the presence of sub-µm sized gaps and pores are detected supporting the findings on the porosification process, drawn from Fig. 6b. Again, the results are similar to those shown in Fig. 6c when increasing on this high temperature level t to 8 h, besides a more pronounced pore formation and a larger penetration depth which can not be fully exploited via FIB technique due to a depth range for characterization limited to about 20 µm. From these morphological investigations it can be concluded that the grain-near portion of the glass matrix is very important to enable the penetration of the phosphoric acid into the depth. This selective etching behaviour may be either based on the microstructure of the glassy envelope close to the Al_2O_3 particles having a lower degree of crystallinity (e.g. amorphous) or on a different chemical composition to the residual glass matrix, as pure SiO_2 has a negligible etch rate in respect to phosphoric acid.

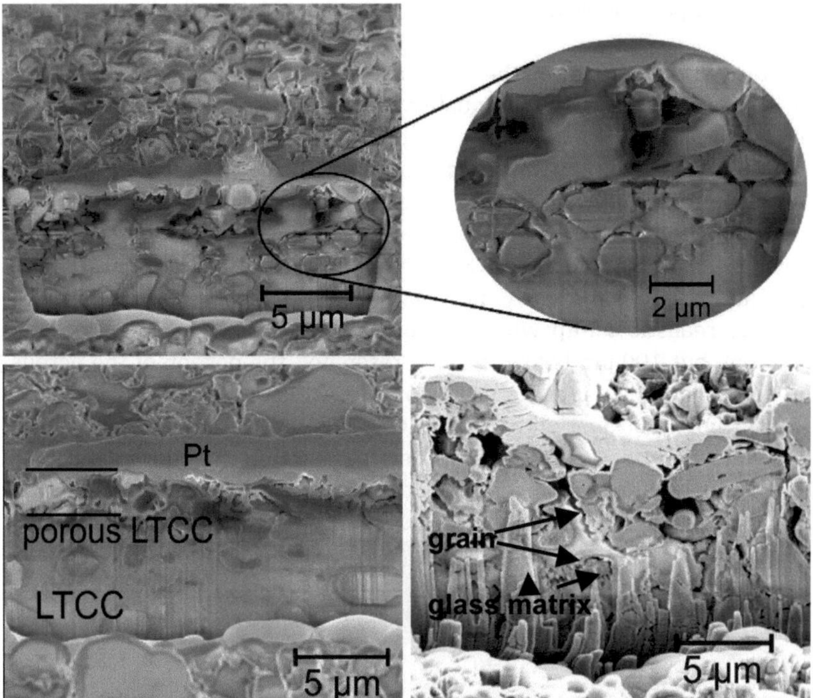

Fig. 6. Cross-sectional view on the microstructure of LTCC close to the surface. The glass-ceramic is etched in phosphoric acid for
a) 8 h at T_b= 90°C (upper row),
b) 2 h at T_b= 110°C (lower left) and for
c) 8 h at T_b= 110°C (lower right).
The platinum top layer is only applied to avoid any structural damage to the probe surface during FIB preparation procedure.

From the results presented so far, an increase in surface roughness is expected when generating a higher degree of porosification or a larger penetration depth. Starting from the values associated with the "as manufactured" LTCC substrate, R_a stays below 1 µm, whereas the maximum surface roughness can range up to several microns. Despite the increase in surface roughness conventional thick film metallization including post firing levels around 850°C can be applied similar to "as manufactured" LTCC substrates without damaging the porous structure.

To measure the change in permittivity ring resonators are arranged on porous LTCC substrates. All analyses were done on DP951 LTCC ceramics with a substrate thickness of d_{LTCC}= 820 µm. The permittivity is determined from a resonance peak at around 6 GHz.

In Tab. 1, the impact of the porosification depth on the dielectric constant is shown. Starting with the quantity $\varepsilon_{r,eff}$ directly determined from the ring resonator the influence of the air is eliminated according to Eqs. 3a, 3b and 4 resulting in the modified values labelled with ε_r. As expected, a substantial, but not very effective reduction in permittivity is achieved when increasing d_p up to 35 µm independent of the evaluation stage. A value of 7.8 is determined for the "as manufactured" LTCC fitting very well into data published recently [13] and giving confidence into the proposed procedure. Assuming, however, a more realistic set-up when placing the ground plane not in a distance of 820 µm, but 100 µm below the ring resonator, the portion of the porous layer in respect to d_{LTCC} is more pronounced. Calculations show that the effective permittivity of a layer with d_{LTCC}= 100 µm is below a value of 3 at a maximum porosification depth of 35 µm. Under these conditions, the modified LTCC is competitive with organic materials commercially available on the market.

Porosification depth d_p	$\varepsilon_{r,eff}$	ε_r	*Calculated value for d_{LTCC}= 100 µm* $\varepsilon_{r,100\,\mu m}$
0 µm	*5,20*	*7,8*	*7,8*
2 µm	*5,12*	*7,22*	*4,69*
15 µm	*5,10*	*7,19*	*4,54*
35 µm	*4,62*	*6,45*	*2,87*

Tab. 1. Measured permittivity for various porousification depth of LTCC. ε_r is the calculated permittivity of the 820 µm LTCC, ε_r ,100 µm is the calculated permittivity for a single LTCC layer of 100 µm

4 Conclusions and Outlook

In this study, a novel process based on a wet chemical etchant is introduced to generate a tailored porosity in fired LTCC substrates (DP 951). Due to the use of a phosphoric acid which is well-known for the patterning of aluminium thin films in MEMS or microelectronic industry a local porosification is feasible by using a photosensitive polyimide as mask material. The method is very simple to monitor, as important parameters, such as the bath temperature and the exposure time, strongly determine the degree and the depth of the porosification process. In addition, the LTCC fabrication process does not need to be changed, as commercially available tapes can be used. Typically, this process step could be implemented in the flow chart of a device or module before final metallization. It could be demonstrated that at a bath temperature of 130°C, the penetration depth has a maximum value of about 40 µm. To enable the penetration of the wet etchant into the LTCC body, this portion of the glass matrix enveloping the Al_2O_3 grains plays an important role, as it is preferentially etched at the onset of the process. With increasing exposure time, however, the etching of the Al_2O_3 grains becomes more pronounced, so that the originally gap-sized microstructure changes to a pore-sized geometry. Basically, the porosification process is reaction-limited at the beginning, while changing to a diffusion-controlled regime after an exposure time of about 5 h independent of bath temperature and indicated by an almost constant saturation level. Associated with the enhanced porosity is an increase in surface roughness. Therefore, R_a changes from about 0.4 µm in the "as manufactured" state to a maximum value of about 1 µm for the porosification parameters investigated. Finally, test patterns were applied on the modified LTCC substrates using thick film screen printing technique and post-firing temperature up to 850°C. Thus, the standard metallization approach for this glass-ceramic material can be used as in the "as manufactured" state. First measurements of the permittivity indicate a clear reduction with increasing porosification depth. Minimum values for the permittivity in the range of 3 using an optimized design make this approach attractive for high frequency applications.

In the near future, further investigations have to be performed to understand in more detail why the glass matrix enveloping the Al_2O_3 grain is etched first and why this feature is based on the corresponding microstructure or on the chemical composition. Due to the challenges linked with the determination of quaternary phases it is assumed that a chemical analysis of the elements dissolved from the LTCC in the etchand will give more insight in the mechanisms involved. Furthermore, the impact of the porosification on the dielectric properties, such as the permittivity and dielectric losses, has to be determined in more detail by varying in particular the bath temperature and the exposure time to exploit the full potential of this promising procedure.

Acknowledgements

This work was performed within the RADARAUGE project (http://www.radarauge-project.com/) financially supported by the German Federal Ministry of Education and Research (BMBF) under contract number 16SV2080. This support is gratefully acknowledged.

References

[1] His, C. S., Hsieh, F. M. & Chen H. P., Characteristics of thick film resistors embedded in low temperature co-fired ceramic (LTCC) substrates, J. Eur. Ceram. Soc., 2007, 27, 2779-2784.

[2] Jantunen, H., Kangasvieri, T., Vähäkangas, J,. and Leppävuori, S., Design aspects of microwave components with LTCC technique, J. of Eur. Ceram. Soc., 2003, 23, 2541-2548.

[3] Fonseca, M. A., English, J. M., Arx, M. V. and Allen, M. G, Wireless micromachined ceramic pressure sensor for high temperature applications. J. Microelectromech. Sys., 2002, 11, 337-443.

[4] Gongora-Rubio, M. R., Espinoza-Vaaejobs, P., Sola-Laguna, L., and Santiago-Avilés J. J., Overview of low temperature co-fired ceramics tape technology for meso-system technology (MsST), Sens. Actuators A, 2001, 89, 222-241.

[5] Ohnuki, Y., Ori, T., Yoshihare, K., Nonaka, Y., Senda, M., and Shibuya, H., Development of copper-polyimide thin film multilayer on LTCC substrate, In Proceedings of International Microelectronics 1994 conference (IMC), 1994, Omiya, pp. 215-219.

[6] Dernovsek, O., Preu, G., Wersing, W., Modes, C., Eberstein, M., Schiller, W., Güther, W. and Schulz, B., Glaskeramikmasse und Verwendung der Glaskeramikmasse. DE 10043194 A1, German patent, 2000.

[7] Chen, G.-H., Effect of replacement of MgO by CaO on sintering, crystallization and properties of $MgO\text{-}Al_2O_3\text{-}SiO_2$ system glass-ceramics, J. Mater. Sci., 2007, 72, 7239-7244.

[8] Jain, A., Rogojevic S., Ponoth S., Agarwal, N., Matthew, I., Gill, W. N., Persans, P., Tomozawa, M., Plawsky, J. L., and Simonyi, E., Porous silica materials as low-k dielectrics for electronic and optical Interconnects, Thin Solid Films, 2001, 398-399, 513-522.

[9] DuPont, Product data sheet, 951 Low-Temperature Cofired Dielectric Tape.

[10] Deisinger, U., Stiegelschmitt, A., Roosen, A., Schwanke, D., Bechthold, F., and Schmaus, C., Charakterisierung und Schwindungsverhalten von LTCC-Grünfolien, Plus, 2001, 3, 2-8 (in German).

[11] Zimmermann, D., Mobley, T., Miller, M., Nair, D.l, Walsh, M., and Smith, M., 20 to 90 GHz broadband characterization of LTCC materials for transceiver modules and

integrated antennas, In: Proceedings IMAPS Advanced Technology Workshop on Automotive Microelectronics and Packaging 2007, Dearborn, paper No. WP25.
[12] Madou, M. J., Fundamentals of microfabrication. 2nd Ed., CRC Press LLC, Boca Raton, Fl, 2002, pp 206-220.
[13] R. Kulke, „LTCC - An Introduction and Overview",edited by the IMST GmbH (http://www.ltcc.de or http://www.imst.de).

Achim Bittner, Hartmut Seidel, Ulrich Schmid
Universität des Saarlandes
Campus, Geb. A5.1
66123 Saarbrücken
Germany
a.bittner@lmm.uni-saarland.de
seidel@lmm.uni-saarland.de
u.schmid@lmm.uni-saarland.de

Thomas Haas
Micro Systems Engineering GmbH
Abteilung ENS
Schlegelweg 17
95180 Berg
Germany
thomas.haas@Mse-microelectronics.de

Keywords: porosity, dielectric properties, glass ceramics, substrates, wet etching

Determination of Complex Permittivity of LRR Radome Materials Using a Scalar Quasi-Optical Measurement System

F. Pfeiffer, E. Biebl, Technische Universität München,
K.-H. Siedersberger, AUDI AG

Abstract

We developed a low-cost quasi-optical measurement system to determine the complex permittivity in the E-band (from 60 to 90 GHz). The evaluation is done in a non-destructive way and can be used for all kinds of single-layered and multi-layered dielectric materials. The method is based on measurements of the scalar transmission coefficient through planar samples for several angles of incidence and two different polarization states (parallel and perpendicular to the plane of incidence). A numerical optimization technique is used to derive the complex permittivity from the measured coefficients. The method utilizes a physical model of a dielectric slab of known thickness, which assumes that a plane wave is incident on the surface of the dielectric material. Measurement results are presented, which are in good agreement with data from the literature.

1 Introduction

An increasing number of mid and high class car models are equipped with radar sensors, which use millimetre wave signals. These sensors are usually mounted at highly design relevant positions in the automobile front. Therefore radomes are often used to cover the sensors. The radome's surface generally affects electromagnetic wave propagation. For a proper radome design and an appropriate choice of the material it is essential to know accurate data for the relative permittivity ε_r and loss tangent at the radar's operating frequency. Therefore we developed a scalar free space transmission measurement system operating in the E-Band – appropriate for automotive long-range 76 to 77 GHz radar (LRR) applications. A picture of the system's free space part is shown in Fig. 1.

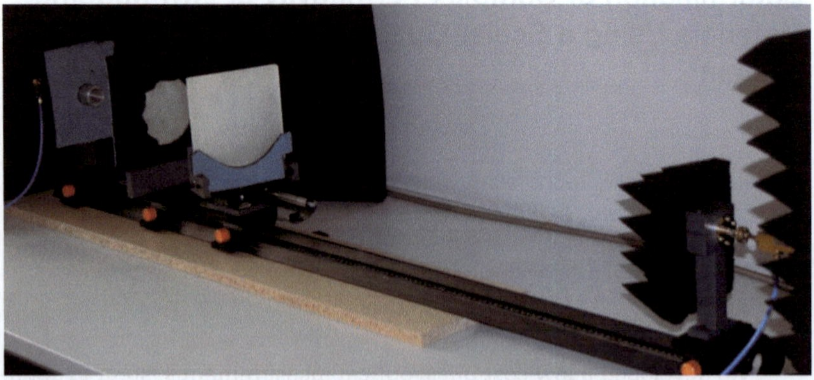

Fig. 1. Picture of the free space measurement system

For millimetre wave material characterization a free space measurement configuration is attractive due to the non-destructive nature of the system and the little effort required for sample preparation. Furthermore, we decided for a scalar measurement system due to the lower cost of the measurement equipment (compared with vector measurement equipment) and the simplicity of the mechanical setup. Free space phase measuring at millimetre wavelengths, would require a highly precise mechanical adjustment in the range of micrometers.

2 Measuring Setup

Based on Gaussian beam optics, we designed the measurement system. A schematic representation of the optical path is shown in Fig. 2.

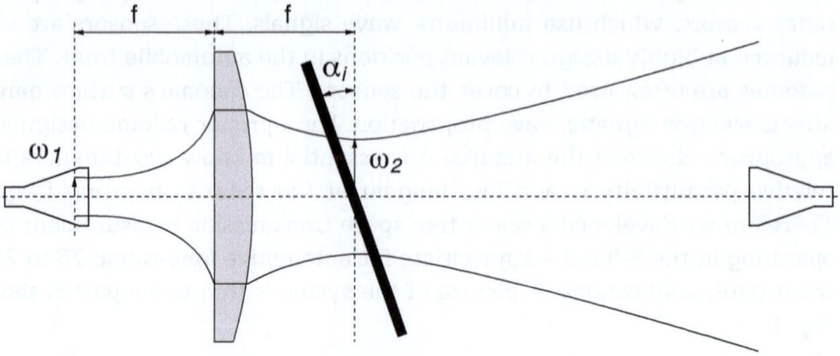

Fig. 2. Schematic representation of the optical path

The system consists of a pair of corrugated horns with Gaussian beam radiating characteristic. The transmitting horn illuminates a focusing lens with a focal distance f of 210 mm. The diameter of the dielectric lens is 131 mm and hence approximately three times the maximum beam waist at this position. Therefore 98.9 % of the beam's power goes through the lens. The sample material is inserted at the output beam waist location, where the wave fronts are flat. In order to minimize the diffraction effects, the size of the dielectric sample is larger than three times the radius of the collimated Gaussian beam w_2 over the entire frequency range. The Gaussian beam is received by an identical horn antenna. The sample holder is mounted on a precision rotary stage to allow a variation of the incident angle α_i. In the receiving path a harmonic mixer is used to down-convert the RF-signal to an intermediate frequency of 741.4 MHz. The power of the down-converted signal is measured with a spectrum analyser R&S FSP40. The scalar transmission coefficient T is calculated by dividing the measured power P_m by the power P_{ref} of a reference measurement without a sample in the free-space path

$$T = \frac{P_m}{P_{ref}} \tag{1}$$

This normalization procedure is the only error correction available for scalar measurement and eliminates only the frequency response. In order to minimize the influence of multiple reflections between sample, transmitting and receiving horns only angles of incidence α_i bigger than 10° are used. The maximum angle of incidence is limited to 40°, to avoid diffractions from the edges of the sample holder and the sample itself.

3 Theoretical Outline

For the determination of the complex permittivity $\varepsilon_r = \varepsilon_r' - j\varepsilon_r''$ a theoretical model for the transmission coefficient $t_{E,H}$ for both polarization states is used.

The model is based on a plane electromagnetic wave incident on an infinite dielectric slab, as shown in Fig. 3.

Since the tangential part of the electric field E and the magnetic strength H must be continuous across the boundaries (air / dielectric slab), the incident field can be calculated in terms of the transmitted field [2]. The transmission coefficient can be obtained by dividing the fields.

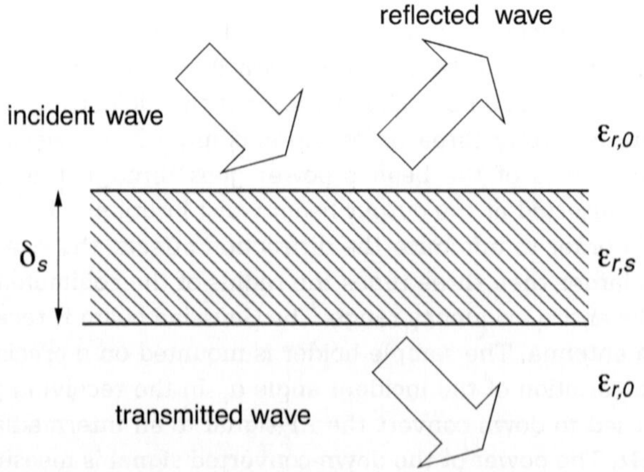

Fig. 3. Model of a plane wave incident on a dielectric slab

The transmission coefficient is then given by

$$t^{E,H} = \frac{E_t^{E,H}}{E_i^{E,H}} = \frac{2Y_0^{E,H}Y_S^{E,H}}{2Y_0^{E,H}Y_S^{E,H}\cos(k_0h_s) + j\sin(k_0h_s)((Y_0^{E,H})^2 + (Y_S^{E,H})^2)} \tag{2}$$

Where Y_i^E is the effective wave admittance for the E polarization

$$Y_i^E = \frac{\cos\alpha_i}{Z_i} = \cos\alpha_i \cdot \sqrt{\frac{\varepsilon_0\varepsilon_{r,i}}{\mu_0\mu_{r,i}}} \tag{3}$$

and Y_i^H is the effective wave admittance for the H polarization.

$$Y_i^H = \frac{1}{Z_i \cdot \cos\alpha_i} = \frac{1}{\cos\alpha_i} \cdot \sqrt{\frac{\varepsilon_0\varepsilon_{r,i}}{\mu_0\mu_{r,i}}} \tag{4}$$

The lower index i indicates the dielectric material. Where S stands for the dielectric sample and 0 for the surrounding air. The product of the wave number k_0 and the effective length h_s represents the phase difference of the plane wave transmitted through the dielectric slab and is given by

$$k_0h_s = k_0d_s\sqrt{\varepsilon_{r,s}\mu_{r,s}}\cos\alpha_s \tag{5}$$

k_0 is the wave number in free-space and d_S the physical thickness of the dielectric sample. The refracted angle α_S in the dielectric material can be calculated from the incident angle α_i using Snell's law. Assuming that the relative permeability is one in the dielectric sample, the transmission coefficient only depends on the complex permittivity and the geometrical measurement conditions. As there is no analytical way to express the complex permittivity in terms of the transmission coefficient $t^{E,H}$, the incident angle α_i and the thickness of the slab d_s, it will be determined by fitting the theoretical model to the measured values of magnitude. For the parameter fitting a non-linear, least-squares method is used. The error function is defined as the sum of the error functions for E and H polarization.

$$EF = EF^E + EF^H = \sum_{k=1}^{N}\left(\left\|t_{c,k}^{E}\right\| - \left\|t_{m,k}^{E}\right\|\right)^2 + \sum_{k=1}^{N}\left(\left\|t_{c,k}^{H}\right\| - \left\|t_{m,k}^{H}\right\|\right)^2 \quad (6)$$

$t_{c,k}^{E,H}$ and $t_{m,k}^{E,H}$ are the calculated and measured transmission coefficients for N angles of incidence. To minimize the non-linear error function a Gauss-Newton algorithm is used.

4 Results

The measurement system has been tested with a dielectric sample of REXOLITE® material with a thickness of (12.90±0.02) mm. The sample thickness has been measured mechanically with a measuring gauge.

In a first step the scalar transmission coefficient has been measured at 201 different frequencies between 60.1 GHz and 89.9 GHz and an angle of incidence of 10°. A coarse approximation of the complex permittivity has been calculated with the assumption of constant permittivity over the entire frequency range. In a second step the measurement has been implemented for 65, 75 and 85 GHz for 62 incidence angles in the range of $10° \leq \alpha_i \leq 40°$, respectively. The complex permittivity has been calculated numerically using as starting value the result of the first step.

As shown in table 1, the results are in good agreement with the complex permittivity, we presented some years ago measured with a high accurate vectorial W-band (75 to 95 GHz) measuring system [1].

	ε'_r	$\tan\delta$	ε'_r	$\tan\delta$
f in [GHz]	*Scalar meat.*		*Vectorial meas. [1]*	
65	2.526	$0.1\cdot10^{-3}$	-	-
75	2.528	10^{-3}	2.5294	10^{-4}
85	2.531	10^{-3}	2.5293	10^{-4}

Tab. 1. Complex permittivity of REXOLITE® at 65, 75 and 85 GHz

Some further measurements have been made to characterize polypropylene, polyethylene, Teflon, PVC, Makroblend® - a PC / PET based polymer used at AUDI AG for LRR radomes. The following table shows the dielectric constant and loss tangent at the LRR's operating frequency of 76.5 GHz.

Material	ε'_r	$\tan\delta$
Polypropylene	2.25	$0.4\cdot10^{-3}$
Polyethylene	2.32	$0.3\cdot10^{-3}$
Teflon	2.05	0
PVC	2.89	$8.3\cdot10^{-3}$
MAKROBLEND®	2.81	$9.6\cdot10^{-3}$

Tab. 2. Complex permittivity at 76.5 GHz

We plan to extend the method in order to characterize layered and composite materials that are usually used as bumper and front grill.

References

[1] G. Friedsam. Bestimmung der komplexen Permittivität und Permeabilität im Millimeterwellenbereich, PhD thesis Technische Universität München, 1998.

[2] E. Hecht, Optik. Bonn/München, Germany: Addison-Wesley, 1989.

Florian Pfeiffer, Erwin Biebl
Technische Universität München
Fachgebiet Höchstfrequenztechnik
Arcisstr. 21
80333 München
Germany
Pfeiffer@tum.de
Biebl@tum.de

Karl-Heinz Siedersberger
AUDI AG
85045 Ingolstadt
Germany
Karl-Heinz.Siedersberger@AUDI.de

Keywords: complex permittivity, quasi-optical measurement, microwave measurement, material characterization

Speed Sensors for Automotive Applications Based on Integrated GMR Technology

K. Kapser, S. Zaruba, Infineon Technologies AG
P. Slama, E. Katzmaier, Infineon Technologies Austria AG

Abstract

Magnetic sensors are widely used in modern cars for example as wheel speed sensors for ABS or to determine crank shaft position in motor management systems. Increasing demand for larger air gaps and better noise performance (lower jitter) requires sensor principles with higher sensitivities compared to the currently used devices. The presented approach with the monolithic integration of GMR (Giant-Magneto-Resistors) sensing elements on top of the signal conditioning ASIC provides a smart, single chip sensor solution which combines superior performance with high reliability.

The presented family of GMR based wheel speed sensors covers a wide range of automotive applications where speed and direction information is required together with precise timing (low jitter) and high air-gap capability. The described sensors can be used with a magnetized pole wheel or in a backbias configuration with a ferromagnetic target wheel.

1 Introduction

A growing demand for sensors in modern vehicles, caused either by safety requirements, comfort features, or legislation drives the development of new measurement principles and sensors with high performance and increased capabilities. To meet the high reliability requirements in the automotive industry, an integration of the sensing element and the signal conditioning electronics within a single chip is advantageous to minimize failures due to interconnections. In addition, this integration gives a small, smart sensor element which can be manufactured in a mainly standard silicon technology with high yield and low cost.

The giant magneto resistive (GMR) effect was discovered in 1988 and shows a resistivity which is depending on the external magnetic field. Compared to other magnetic sensing principles as Hall or AMR (anisotropic magneto resistance), the GMR effect exhibits the largest sensitivity and lowest noise. This makes the technology well suited for automotive applications which require a large air gap and/or a very precise timing (low jitter). The GMR sensing elements can be monolithically integrated in a standard BiCMOS ASIC which contains the signal conditioning circuitry. Based on this concept, a family of sensors is developed which targets applications including wheel speed measurement for the Antilock-Braking Systems (ABS) or crankshaft speed and position measurement for motor management systems.

2 The GMR Effect and the Spin Valve Configuration

The GMR effect can be understood as interface scattering of the conducting electrons when moving through a stack of magnetic layers (typically Ni, Co or Fe) separated by non-magnetic spacer layers. Each electron has a spin which can be either "up" or "down". The scattering probability of the electrons depends on the orientation of the spin and the magnetic moment of the layer. Parallel orientation yields a low scattering probability (low resistance) whereas anti-parallel alignment leads to a high scattering probability and therefore to a high electrical resistivity (Fig .1)

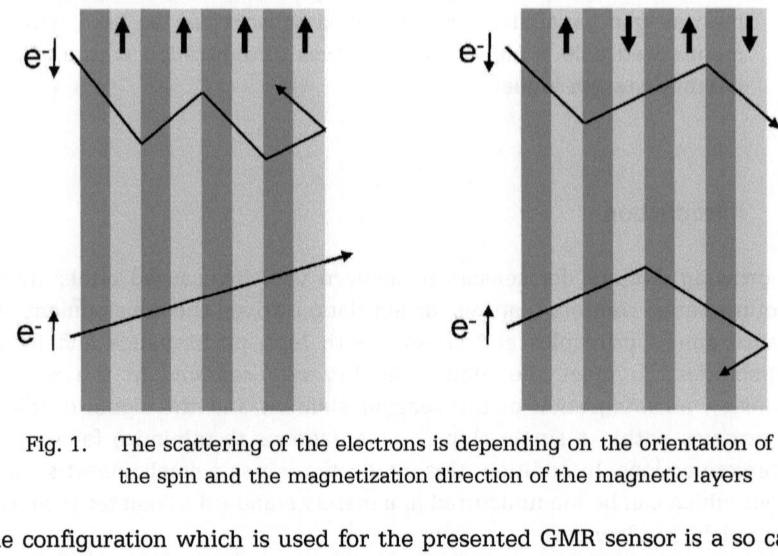

Fig. 1. The scattering of the electrons is depending on the orientation of the spin and the magnetization direction of the magnetic layers

The configuration which is used for the presented GMR sensor is a so called spin valve [1]. This structure consists basically of a "hard" layer, the "pinned" layer, which has a fixed magnetic orientation whose direction cannot be changed by an external magnetic field. This layer is separated and magnetically decoupled by a non-magnetic spacer from the "soft" or "free" layer with a magnetization which can perfectly follow the external (in plane) magnetic field (Fig. 2).

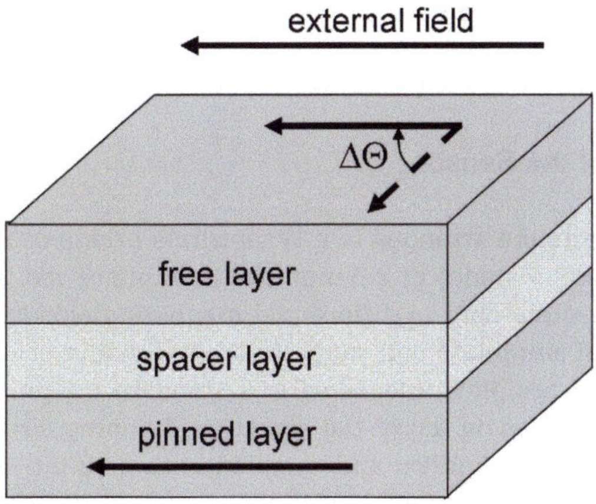

Fig. 2. Spin valve structure with pinned layer and free layer.

The electrical resistivity of such a structure depends on the orientation of the magnetization of the pinned layer with respect to the free layer, whose orientation can be changed from parallel to anti-parallel applying an external magnetic field. This structure is used as the basic sensing element, the GMR resistor. The electrical characteristic of such a stack is shown in Fig. 3. For magnetic fields exceeding B_k the resistivity saturates. In this case, the magnetization of the pinned layer and the free layer are completely aligned in parallel. Further increase of the external field results in no change of the resistivity. For magnetic fields in the range $-B_k < B < +B_k$ the resistivity shows a linear behaviour.

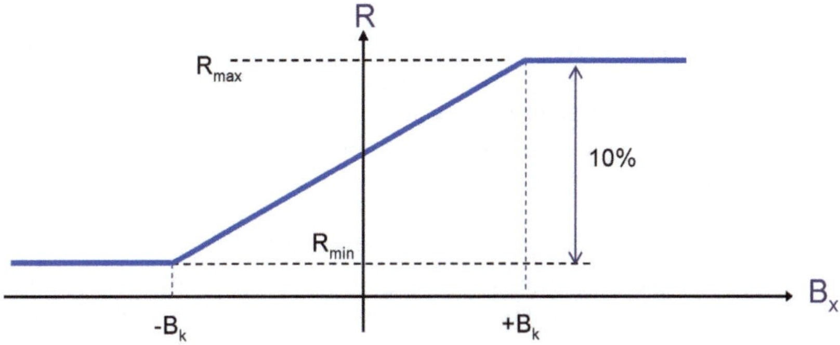

Fig. 3. Electrical characteristic of a spin valve

For a typical spin valve B_k is in the range of 5 mT and the change of resistivity is around 10%. The magnetization direction of the pinned layer of such a spin valve is fixed by coupling to an artificial antiferromagnet, which is also part of the spin valve [2]. Therefore, no external stabilization magnet is necessary in contrast to an AMR based device.

3 Concept of the Sensor

Four GMR resistors are arranged in a Wheatstone bridge configuration on the chip with a spatial distance of 2.5 mm. This comprises the basic sensing element and is sensitive only to differential magnetic fields in x-direction thus increasing signal amplitude and suppressing external interferences and magnetic offsets. The bridge is integrated in a standard 0.5 µm BiCMOS process as last processing step on top of the signal conditioning circuitry. The resulting bridge voltage is amplified and sampled with a 12 bit AD converter and 200 kHz sampling rate. All further signal processing such as offset correction and calculation of the switching thresholds is done in the digital domain. Due to the intrinsic signal clamping of the GMR sensing element (saturation) the input stage of the AD converter can perfectly be adjusted to the maximum possible signal level. In this way there is no need for any programmable gain amplifier (PGA) for signal matching. To determine the optimum switching threshold, which is the zero-crossing of the differential magnetic field, the signal extrema (minima and maxima) are continuously detected by the digital circuitry and the signal offset is calculated. With this information, the zero-crossing of the offset-corrected magnetic signal can be detected and the output of the sensor is switched. The output signal of the IC can either consist of a two wire 7 mA/14 mA current interface for the ABS sensor type or an open collector voltage interface for the crankshaft sensor. The principle block diagram for the ABS sensor is shown in Fig. 4.

To improve noise performance, an internal threshold ΔB_{limit} is defined. The signal maxima have to exceed this value to be recognized as a valid signal. All magnetic fields below that threshold are not considered for evaluation. A zero-crossing of such a signal therefore does not lead to an output switching. This concept is known as hidden hysteresis. As an example, if a vibration modulates the distance of the sensor to the pole wheel, this signal change is suppressed as long as it is below the implemented hysteresis. On the other hand, the value of ΔB_{limit} determines the maximum air gap which can be reached in the application, so that there is a trade off between maximum air gap and immunity to interference. For the discussed device ΔB_{limit} can be selected to be 300 µT for the standard version and ΔB_{limit}= 150 µT for the high sensitivity device. In the case of a large magnetic field however, there is no need to have a small hysteresis. To account for such a condition the hysteresis is adaptive. This means that after power on or chip reset, it is always 300 µT or 150 µT, respectively. As soon as the magnetic signal is tracked by the IC, the hysteresis adjusts to 25% of the present signal maximum.

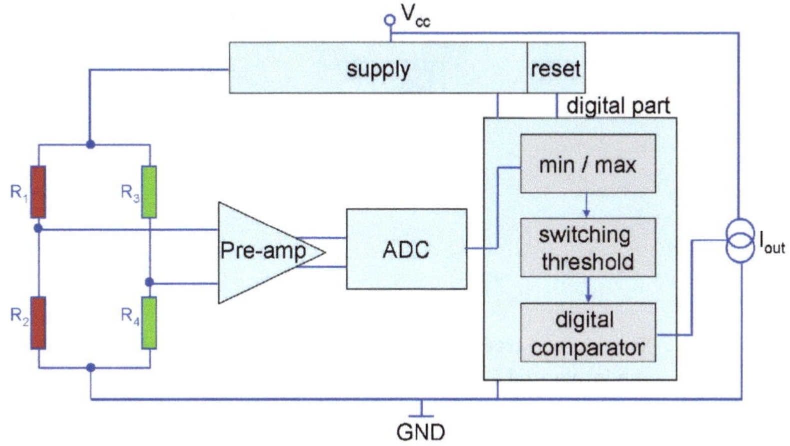

Fig. 4. Block diagram for the integrated GMR (iGMR) wheel speed sensor

Therefore, the noise performance is optimized without sacrificing air gap capability. Only when it is required due to the magnetic circuit or after power on, the sensor operates in the highest sensitivity mode (smallest possible ΔB_{limit}). After power on, where there is no information about the signal maximum and minimum values, the self-calibration algorithm can not work. As the offset is unknown, no switching at zero-crossing can be performed. In this case, the switching threshold is an internal value defined as 2 x ΔB_{limit} (600 µT and 300 µT respectively). For the first and second detected magnetic edge this behaviour is responsible for the output switching in a mode referred to as the uncalibrated mode. Due to this uncalibrated mode of operation, an output signal can be obtained even for the first valid magnetic edge. After two magnetic edges the value of the minimum and maximum magnetic field can be determined and the offset calculated. From there on, the output switching occurs at the true zero-crossing of the magnetic field and the sensor is now in calibrated mode. An additional feature is the true zero Hertz capability of the device. Since there is no reset timer resetting the chip after a certain time without detecting a signal change, it is possible to measure magnetic signal frequencies down to practically 0 Hz. The ABS sensor type has a two pin package with an integrated 1.8 nF capacitor on the leadframe to enhance EMC (electromagnetic compatibility) performance. The crankshaft sensor is packaged in a three pin package with one integrated 47 nF capacitor between VDD and GND and one 4.7 nF capacitors between OUT and GND. (Fig. 5).

Fig. 5. Two pin and tree pin leaded package with integrated capacitor for the integrated GMR speed sensors

4 GMR Sensor for ABS and Crank Shaft Position

The wheel speed information is used for various systems in a vehicle. Well-known is the ABS system which avoids a blocking of the wheel during braking to prevent the car on the one hand from sliding and on the other hand to reduce the braking distance. The ABS functionality is usually a sub-system of the ESC system (electronic stability control). A magnetized encoder is mounted at the wheel hub and the wheel speed sensor measures the magnetic field. To generate a differential magnetic field, the pitch of the pole wheel has to match with the pitch of the Wheatstone bridge on the sensor. This means that for an optimum signal the pole wheel pitch should be close to 5 mm (Fig.6).

The direction of the tangential magnetic field components of the pole wheel determines the direction of the magnetization of the free layer. In the case that one side of the bridge is in the low ohmic condition (free layer parallel to pinned layer), the other side has a high electrical resistivity (free layer and pinned layer anti-parallel). An output signal of the bridge is generated which is evaluated by the signal conditioning circuitry. The GMR sensor is sensitive to the tangential (in-plane) magnetic field components, whereas a Hall sensor is sensitive to the perpendicular field component. As a magnetic encoder wheel has both components in similar magnitude, the identical pole wheels can be used without any change for Hall and GMR based devices. The only difference to Hall is a 90° phase shift of the output switching regarding the mechanical encoder position. This refers to the fact that the zero-crossing of tangential and perpendicular field components is shifted by 90°.

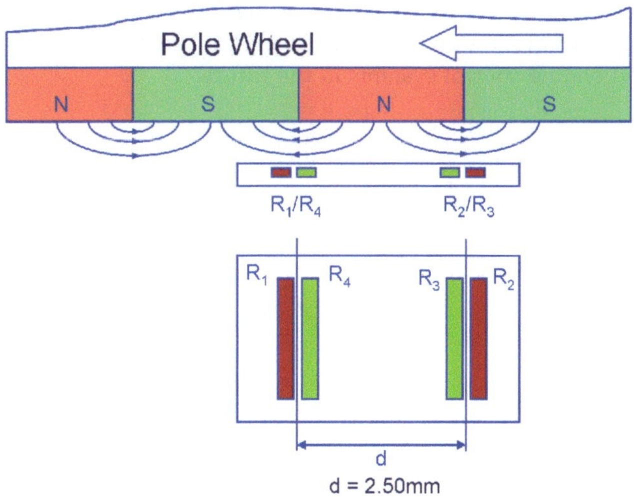

Fig. 6. GMR speed sensor with magnetized pole wheel

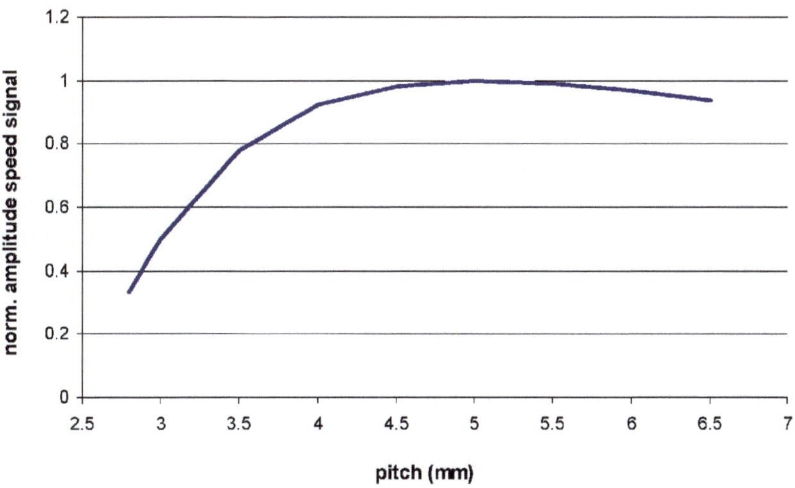

Fig. 7. Relative differential signal amplitude for pole wheels with different pitches

The differential GMR sensor can be used with a wide range of pole wheels with different pitches. A deviation from the optimum pitch value of 5 mm leads only to a small degradation of the signal amplitude (Fig. 7)

The ongoing challenge for improved driving performance, lower fuel consumption and lower emissions goes hand in hand with upcoming lower exhaust and fuel consumption limits set by law. This bears the need for even more advanced engine control systems in the future.

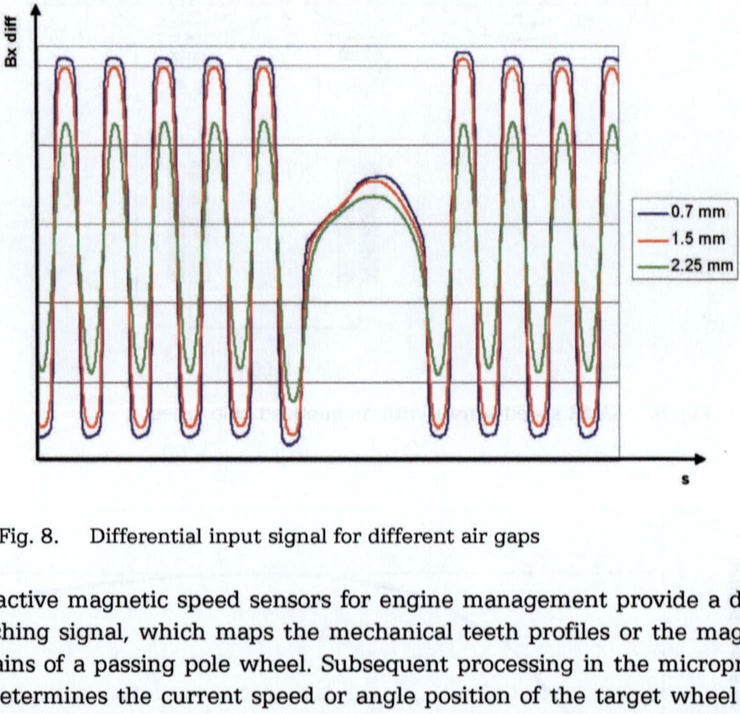

Fig. 8. Differential input signal for different air gaps

The active magnetic speed sensors for engine management provide a digital switching signal, which maps the mechanical teeth profiles or the magnetic domains of a passing pole wheel. Subsequent processing in the microprocessor determines the current speed or angle position of the target wheel from this switching signal. This data is further processed for ignition control and misfire detection.

The task of the sensor is therefore to provide the switching flanks with good reproducibility and high angle accuracy relative to the target wheel. Nevertheless the mounting tolerance of the sensor should be relatively large to keep the system costs low. Summarized, the sensor should provide a phase accurate output signal for magnetic input frequencies from 0 to 10 kHz and over an amplitude range of approximately 1 to 100. The required temperature range is from -40°C to 150°C with an air gap range of 0 to 3.5 mm [3].

In general, pole wheels or gear wheels that are used in crankshaft applications have markings, which affect the magnetic input signal for the sensor near the signature region. This discontinuity also changes the shape of the magnetic signal over air gap (distance sensor to encoder). This can lead to wrong output signals if not considered The complex calibration algorithm has to cope with these demands.

The general function of an active crankshaft sensor is detecting speed, position and rotational direction of the crankshaft.

The direction information is transferred via a 45/90 µs PWM (pulse width modulation) protocol. The frequency of the output pulses corresponds to the rotational speed of the target wheel. The pulse width includes the rotational direction information. During forward rotational direction of the crankshaft, 45 µs pulses are being issued at every magnetic rising edge of the input signal (Fig. 9).

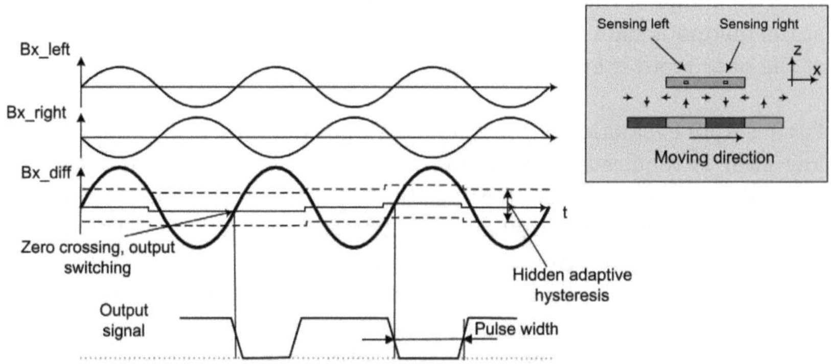

Fig. 9. Differential signal processing, PWM protocol

The rotational direction information is mandatory for new hybrid and start/stop mode engines. A sensor in the center of the IC is used to provide a typically 90° phase shifted signal (Fig. 10).

Depending upon the rotation direction of the target wheel, the signal of the center probe anticipates or lags behind for 90°. This phase relationship can be evaluated and converted into rotation direction information by sampling the signal of the center probe in the proximity of the zero crossing of the "speed" bridge signal. The evaluation of the rotation direction is interesting only at low rotation speed, since a direction of rotation reversal can take place only there. Hence at low rotation the already existing ADC can be used in time multiplexing to convert also the "direction" signal of the center probe without losses of phase accuracy in the output signal [3].

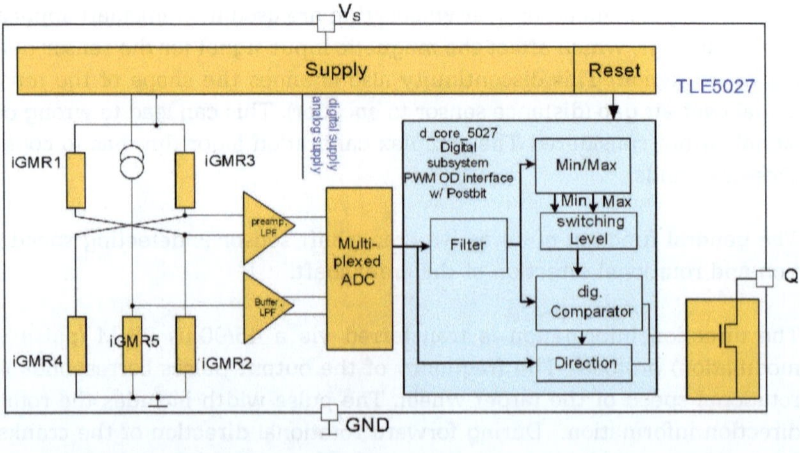

Fig. 10. Block diagram crankshaft sensor

After a change of the rotational direction a 90 µs output pulse is being issued at the magnetic falling edge. In this way the reference to the mechanical edge of the gear tooth is still consistent.

For accurate evaluation of the crank shaft position, a sensor output signal with low jitter is required, thus minimizing phase errors between output switching and mechanical crank shaft position. Due to the high sensor sensitivity of the GMR elements an excellent noise performance is found. The dominating noise source is the digital signal conditioning, e.g. clock noise and quantization noise and not the sensing element itself. As these effects are temperature independent, the measured jitter is mainly constant in the complete operating temperature range. Comparing these results with a Hall based sensor clearly shows the high jitter performance of the GMR based devices (Fig. 11).

A minor dependence on the temperature can be seen only in the case of a very small magnetic field of 0.6 mT. This is a result of the temperature induced noise of the sensing element (GMR resistor), which is the dominant effect for the Hall based device.

Another advantage in magnetic encoder applications which can be achieved with the help of iGMR technology is to place the device parallel to the x/z plane (Fig. 12). This can be done as the GMR sensor is sensitive to the in-plane magnetic field components. With this new arrangement both sensor module size and cost can be reduced. For Hall based sensors it is mandatory to bend the IC at 90° (device parallel to x/y plane) as the sensitive direction is in the z-axis.

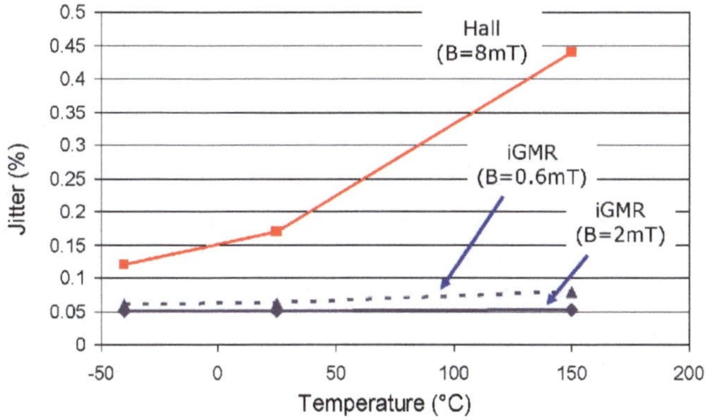

Fig. 11. Jitter performance of a GMR based sensor compared to a Hall device

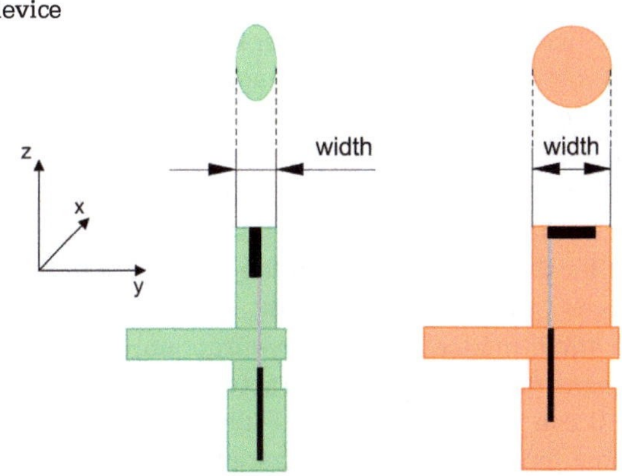

Fig. 12. Left: Slim iGMR sensor module; right: Hall based sensor module

The advantages of the GMR based speed sensor family can be summarized as follows:

- large air gap performance due to the high sensitivity of the sensing element
- low jitter, thus giving high accuracy and small phase error for position detection in crank shaft sensing
- temperature independent jitter even for very small magnetic field values for high accuracy position measurement
- monolithic integration of sensing element and signal condition circuit; additional capacitor integrated on the lead frame of the package to enhance EMC performance
- more flexibility with regard to sensor module design (slim sensor module)

5 GMR Sensor with Backbias Magnet

The use of a GMR speed sensor with a magnetized pole wheel is straightforward as shown in the previous chapters. Some applications, however, require a configuration with a ferromagnetic tooth wheel instead of a magnetized encoder. In this case a magnet has to be attached to the sensor to generate a magnetic bias field ("backbias magnet"). Due to the high sensitivity of the GMR device care has to be taken in this configuration that the backbias magnet does not drive the GMR sensing elements into saturation. Ideally, the bias field has a pure z-component, i.e. pointing perpendicular to the sensor surface. Magnetic fields in this direction have no effect on the GMR resistivity as the field is perpendicular to the sensitive axis of the sensor (x-axis). A ferromagnetic tooth passing by would bend the magnetic flux lines and an in plane field component (x-component) is generated. This is measured by the sensor and the output switching is similar to the case of a magnetized pole wheel. This ideal condition, however, with a pure z-component of the bias field can not be realized in practice. Due to position tolerances of magnet and sensor and the magnetic field distribution of a small magnet, there is always an x-field component present, even without ferromagnetic target wheel. This x-component generates a magnetic offset which brings the GMR resistors into saturation in the worst case and no measurement would be possible. A special magnetic circuit has to be developed, to avoid this magnetic offset and maintain the working point for the GMR element in the center of the resistivity curve. Two possible concepts can solve this problem.

- With a static magnetic field in y-direction (in-plane but perpendicular to the sensitivity axis) the characteristics of the GMR resistor shown in Fig. 3 can be changed. These magnetic field component acts as an additional force on the magnetization of the free layer. A larger external field is required to align this magnetization parallel or anti-parallel to the pinned layer. Therefore, the linear region of the characteristic is extended. A small magnetic offset in sensitivity direction, caused by an inhomogeneous magnet or position tolerances has only a minor influence and causes only a small shift of the working point.
- A special magnet is designed which generates no magnetic offset at the location of the GMR resistor. The magnetization of such a magnet is not homogeneous but is inclined symmetrically in the +x and –x direction.

In Fig. 13 the above mentioned magnets are shown. For the first option (magnet A), the magnetization is inclined in the y direction (not pure z-component). In that way a static field component in y-direction is generated and the linear region of the GMR element is extended. For the magnet displayed on the right side (magnet B) in Fig. 13 the inhomogeneous magnetization in the x-z plane is shown, the size of the magnet in x/y direction being 10 mm x 10 mm. For reference measurements, a magnet with identical size and magnetization but with 0° inclination is used.

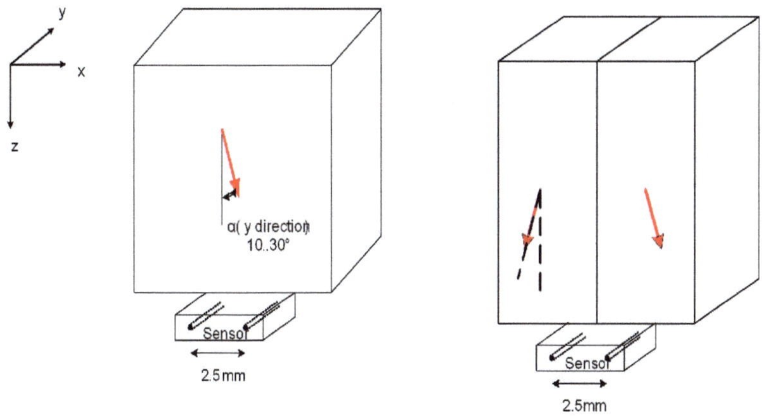

Fig. 13. left: Magnet A with magnetization in z-y plane to increase the linear region of the GMR element. Right: magnet B with inhomogeneous magnetization to generate an offset-free magnetic field.

To evaluate the performance of each of this magnets, the GMR characteristics and the Wheatstone bridge offset in the presence of a rotating ferromagnetic target wheel was measured for different air gaps. The signal offset is calculated from measured signal maxima and minima. Fig. 14 shows the results for magnet A with 20° inclined magnetization compared to the reference magnet with 0°.

It can be seen that the GMR characteristic is changed and the linear region is enhanced from ± 5 mT to more than ± 25 mT. The measured bridge offset for the reference magnet is close to the saturation value of 800 mV. For small air gaps, an almost offset-free signal can be found. The zero-offset signal corresponds to an output value of 1250 mV (center of the GMR characteristics). Therefore, a 0° magnetized magnet is not suitable for a back biased GMR application. For the magnet A, the characteristic has a much large linear range and small signal offsets are therefore not that critical. In addition to that, the remaining offset is smaller and closer to the center of the characteristic. In Fig. 15 the results for magnet B are shown. As there is no magnetic field component in y-direction, the GMR characteristic is unchanged with respect to the reference magnet characteristic. However, the signal offset of magnet B is much closer to the center of the characteristics and exhibits almost no remaining offset.

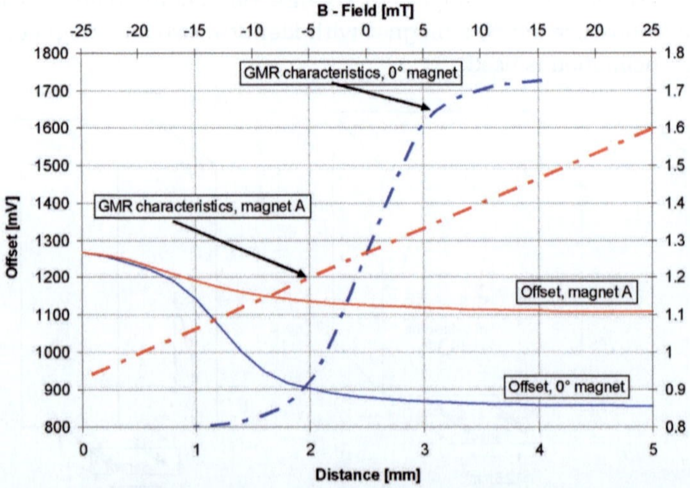

Fig. 14. Measurements of the GMR characteristic and bridge offset of a backbiased sensor for magnet A with 20° magnetization compared to the reference magnet

These results show that with a suitable magnetic circuit design, the GMR speed sensors can be used for applications with ferromagnetic target wheels.

6 Lifetime Measurements

Compliance with qualification requirements including temperatures up to 150°C and lifetimes exceeding 10.000 hours operating life are essential for the use of such a sensor in an automotive application. For an ABS sensor the operating temperature can have peak values up to 190°C, due to the vicinity to the braking disk. Measurements were performed to investigate the stability of the GMR sensor over time and temperature. For this purpose, several samples were stored at T = 250°C for various times and the GMR effect (change of resistivity between high and low saturation value) was measured. Two different etching processes (process A and B) were used for structuring the GMR resistor. A clear dependency of performance degradation on the on etching process was found (Fig. 16).

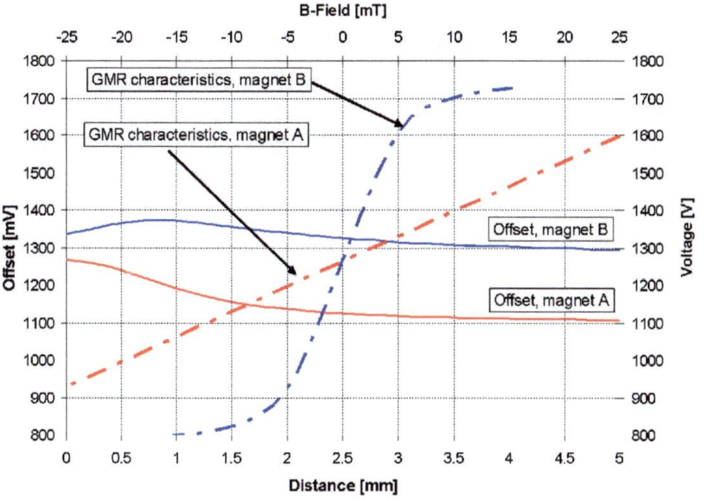

Fig. 15. Measurements of the GMR characteristic and bridge offset of a backbiased sensor for magnet B compared to magnet A

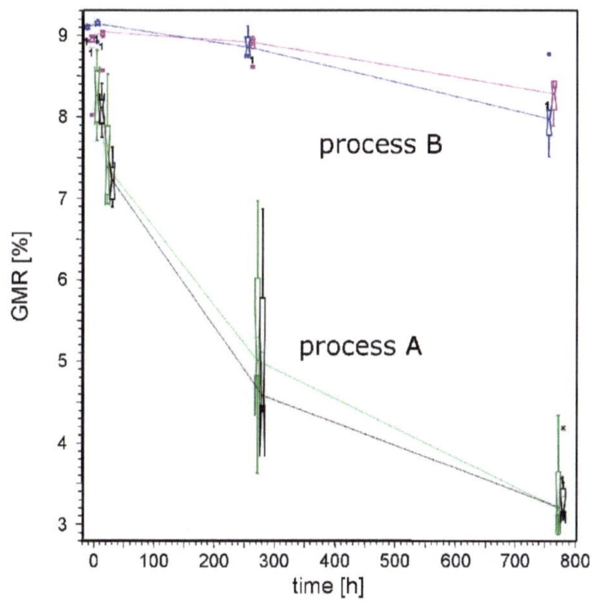

Fig. 16. High temperature storage at 250°C of GMR sensors structured with different etching process A and B.

The main reason for this behaviour is a corrosion effect which occurs at the GMR layers. An improved etching process (process B) was developed which greatly enhanced the stability and lifetime of the sensor. From the measurement data an activation energy of approx. 1 eV for the degradation process was derived. Taking the relative change of 2% of the GMR effect (e.g. resistivity change from 10% to 9.8%) as performance limit, the stability at different operating temperatures can be calculated based on an Arrhenius model. For etching process A this yields a lifetime of more than 400 h at 175°C whereas the limitation for process B is in the range of 10.000 hours at a temperature of 175°C.

7 Summary

In conclusion, we present a family of novel speeds sensor for automotive applications based on an integrated GMR technology with the sensing element and signal conditioning circuit on one IC. Sophisticated calibration and an offset compensation algorithm based on digital signal processing, together with high sensor sensitivity, accounts for superior performance compared to Hall sensors. High air gap capability and low, temperature independent jitter makes the devices ideally suitable for ABS/ESC systems and crank shaft sensing with the potential to address further applications (e.g. transmission speed sensing). Investigations on GMR stability prove that the technology meets the requirements of the automotive industry regarding reliability and lifetime. The sensors can be used for magnetic pole wheel application as well as for ferromagnetic target wheels with a backbias magnet.

References

[1] B. Dieny, VS. Speriosu, S.S.P. Parkin, B.A. Gurney, D.R. Whilhoit, D. Mauri; Phys. Rev.B. 43 (1991)1297-300.

[2] W. Roessler, J. Zimmer, Th. Bever, K. Pruegl, W. Granig, D. Hammerschmidt, E. Katzmaier, "Integrated Giant Magneto Resistors – a new Sensor technology for Automotive Applications", in J. Valldorf, W. Gessner (Eds.), "Advanced Microsystems for Automotive Applications 2006", Springer 2006, pages 323-342.

[3] E. Katzmaier, W. Raberg, K. Kapser, P. Slama, "Integrierte selbstkalibrierende I-GMR Drehzahlsensoren mit 5 µT Phasengenauigkeit für automobile Anwendungen", Sensoren im Automobil, Expert Verlag, 2007.

Konrad Kapser, Sigmund Zaruba
Infineon Technologies AG
81726 Munich
Germany
konrad.kapser@infineon.com
sigmund.zaruba@infineon.com

Peter Slama, Ernst Katzmaier
Infineon Technologies Austria AG
Siemensstr. 2
9500 Villach
Austria
peter.slama@infineon.com
ernst.katzmaier@infineon.com

Keywords: magnetic sensor, GMR sensor, giant magneto resistor, ABS, crank shaft, wheel speed sensor

Micro-Rotation Angle Sensor with Integrated Hall IC

T. Ina, K. Takeda, Nippon Soken, Inc.
A. Sawada, S. Fukaya, Denso Corp.

Abstract

Given the functional sophistication of today's automotive electronic control systems, the need for higher performance sensors is increasing in various technical fields, including engine and vehicle controls. Various sensors have been developed and installed in automobiles in order to detect the mechanical states such as the moving stroke and rotation angle of automotive parts, both of which are important for accurately monitoring the operation of control equipment and devices. The authors developed a high accuracy micro-rotation angle sensor incorporated with a Hall IC. Because of a newly devised magnetic circuit, this sensor achieves small size and high environmental tolerance, and is already being used in the latest on-vehicle control systems.

1 Background of Sensor Development

1.1 Rotation Angle and Moving Stroke Sensors in Vehicles

In keeping with sophistication and functional improvement of automotive electronic control systems, there has been an increasing demand for sensors that detect the mechanical conditions of control equipment and the operating conditions of drivers in various technical fields including engine and vehicle controls. In this connection, rotation angles and moving strokes are typical physical quantities for determining these conditions. Under such circumstances, as shown in Fig. 1, rotation angle sensors are the main currents. Furthermore, in practice, many sensors mechanically convert moving strokes into rotation angles as a way to determine moving strokes. When detecting such rotation angles, mainstream needs used to be to detect relatively large angles. However, with the need for small size and high accuracy for vehicle equipment, there has been an increasing need for high-performance angle sensors which can detect micro-angles (less than several degrees) with high accuracy and which have environmental resistance, durability, and cost-competitiveness, that are sufficient for on-vehicle sensors.

Among the detection methods capable of meeting such requirements, magnetic sensors excel in terms of cost and durability, but this type of sensor has suffered from insufficient accuracy.

This presented the challenge of producing high-accuracy magnetic sensors, which the authors have solved by devising a simple new magnetic circuit using a Hall IC.

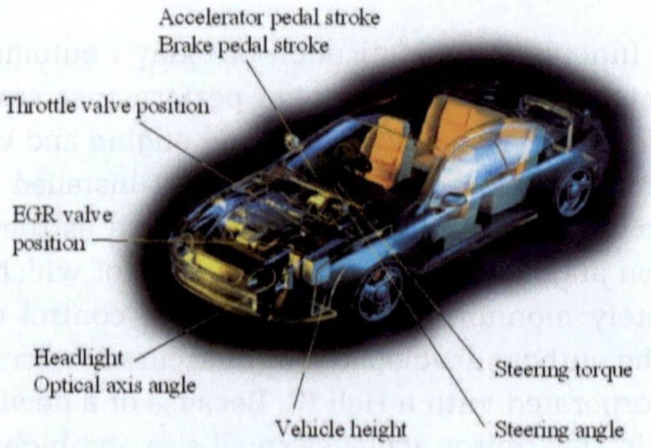

Fig. 1. Typical rotation angle and moving stroke sensors mounted in a vehicle

1.2 Typical Angle Sensors

Fig. 2 shows some structures, output signals, and subjects of typical rotation angle sensors. An optical sensor can detect micro angles if the slit width is narrow, but the costs associated with this are high. Furthermore, it is necessary to provide protection for the sensor and to restrict the use environment when the sensor is to be used in the environment required for automotive sensors (including temperatures and stains). On the other hand, a sensor using Hall or magneto resistive (MR) elements has high environment resistance but has so far not been made sensitive enough to measure micro angle ranges. This fact constituted an obstacle to the achievement of high accuracy.

Type	Structure	Output signal	Subject
Optical type	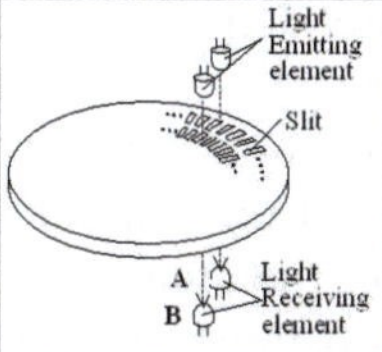	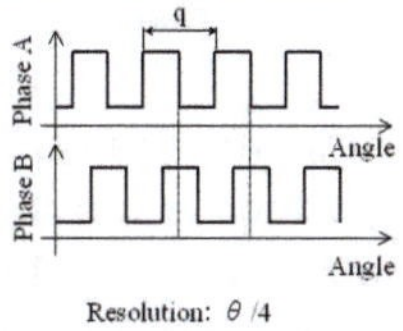	· Expensive · Unusable at high temperature
Magnetic type	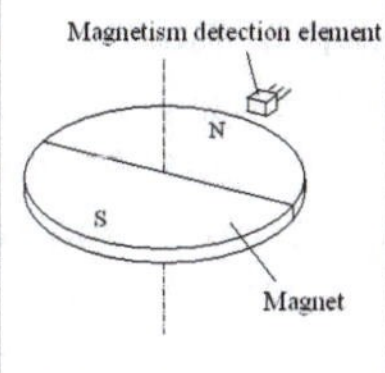	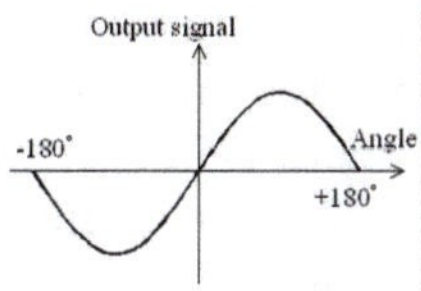	· Poor Sensitivity to angle change

Fig. 2. Typical rotation angle sensor

2 Concept for Improvement of Sensor Sensitivity

Fig. 3 shows the structure of a basic rotation angle sensor. The magnetic detection element used here is a Hall IC, which has a temperature characteristic correction function and which is capable of adjusting output voltages with respect to producted magnetic flux densities. Rotation angles of the magnet, which is a rotor, are detected as variations in the magnetic flux density that passes through the yoke portion (portion A in this figure).

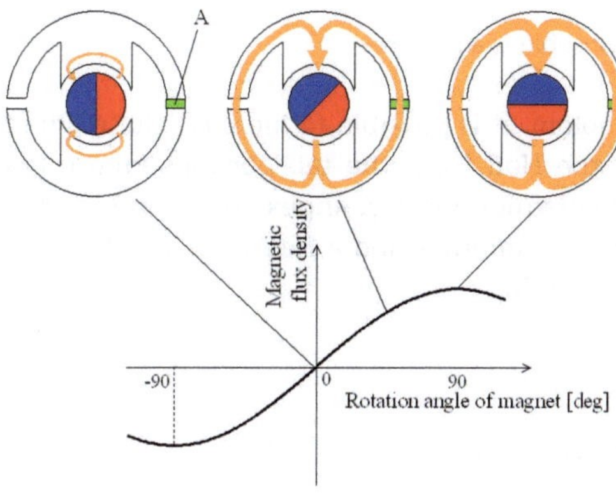

Fig. 3. Basic structure of sensor

Generally speaking, detection accuracy depends on the extent of variations in magnetic flux density in the measurement range. For a sensor with the basic structure in Fig. 3, if, for example, a micro-angle of about 2 to 3 degrees is to be measured, the variation in magnetic flux density decreases significantly.

The rotation angle detection range can be reduced by increasing the number of poles of the magnet, which is a rotor, and the number of yoke tabs. Fig. 4 shows a case where the number of poles of the magnet is six and the number of tabs is six. While this decreases the detection range to one-third of that of a sensor with this basic structure (as shown in Fig. 3), at the same time, the maximum value of magnetic flux density that passes through the Hall IC also decreases. Therefore, there is almost no difference in magnetic flux density variation with a change in angle. In the figure, this is shown by the blue curve. The reason for this is that the magnetic flux that passes through the Hall IC experiences only the magnetic flux from nearby magnetic poles, thus failing to utilize the magnetic flux from more distant poles.

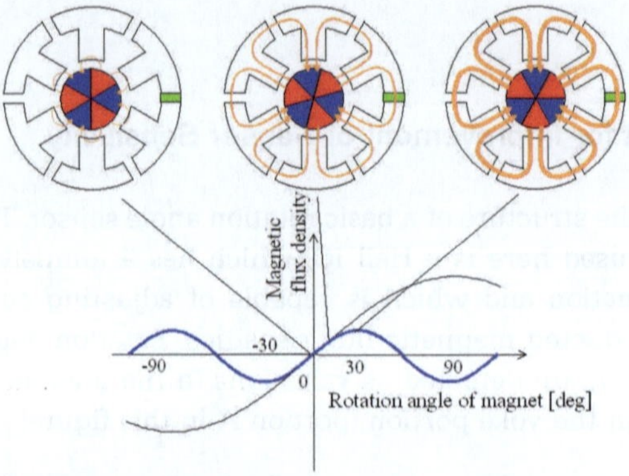

Fig. 4. Multipole magnet case

To solve this problem, it is possible to increase the magnetic flux density that passes through the Hall IC, if the tab portions where magnetic flux density variations due to magnet angle changes are the same are concentrated in one place in some way or another and a Hall IC is installed there, as shown by the red curve in the Fig. 5.

4 Sensor as Devised

To put this concept into practice, we devised a sensor with a new structure shown in Fig. 6. The sensor consists of an annular multipole magnet with alternating N- and S-poles, as well as two yokes on the outer periphery of the multipole magnet, each with a comb-tooth shaped tab. In the initial state, the tab portions are located at the boundary between an N-pole and an S-pole, and the magnetic flux from the magnet flows only between the tabs without passing through the area between the yokes. When the multipole magnet rotates, then what happens, for example, is that a large part of the tab portion of the upper yoke is in contact with an N-pole, while a large part of the tab portion of the lower yoke is in contact with an S-pole, so the magnetic flux from the magnet flows from the upper yoke to the lower yoke. This enables detection of the magnetic flux, since a Hall IC is arranged between the upper and lower yokes.

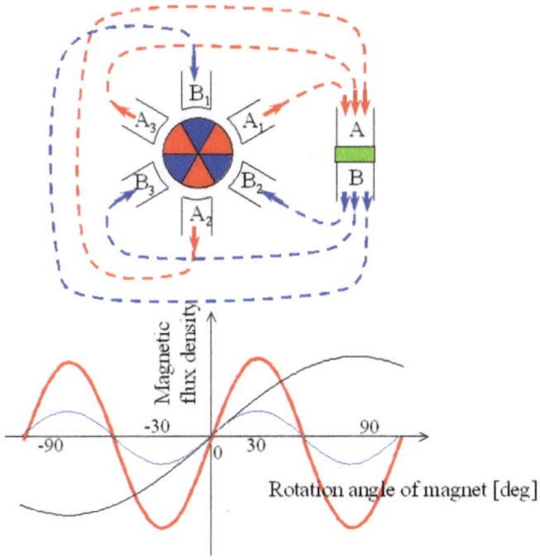

Fig. 5. Method of increasing magnetic flux density variations

As mentioned above, by adopting a multipole magnet it is possible to reduce the angle detection range. Furthermore, the magnetic flux generated by the magnet is concentrated in one place, and a Hall IC is arranged there, with the result that the variation of magnetic flux density with angular change is increased, thus improving detection accuracy. Moreover, the magnetic flux flowing from the magnetic poles to the yoke tabs is concentrated in one place, and therefore, the effects exerted on the polarization scatter of the magnet and on the dimensional scatter of the pork tabs are averaged out, thereby contributing to the improvement of the detection accuracy.

Fig. 7 shows the magnetic flux density that passes through the Hall IC as the number of magnetic poles is changed. It is possible to obtain ideal magnetic flux density variation by selecting the appropriate number of poles for the desired rotation angle range.

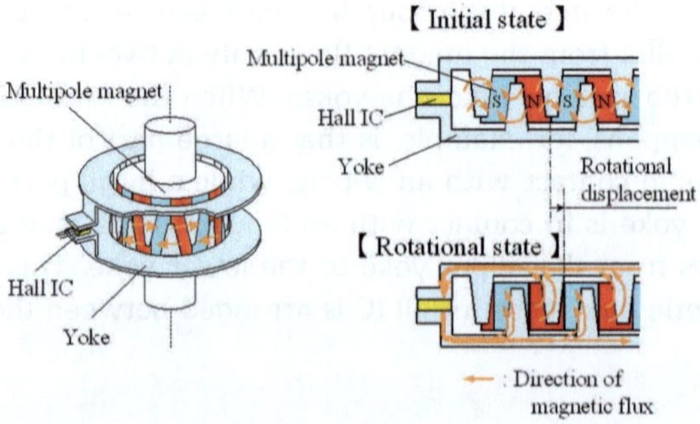

Fig. 6. Explanatory diagrams of sensor structure and operation

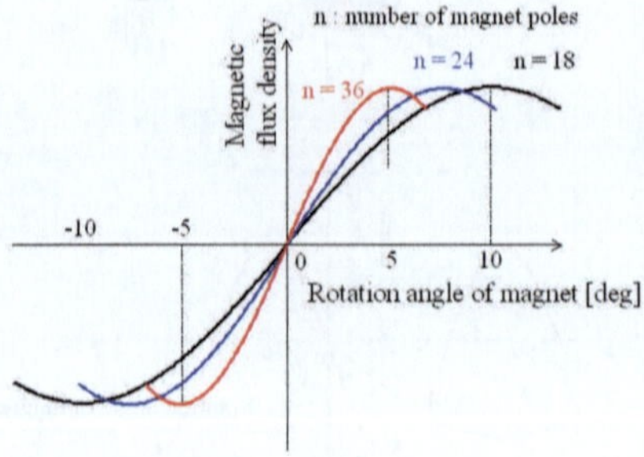

Fig. 7. Sensitivity improvement effect due to adoption of multipole magnet

4 Actually Applied Sensor Structure

Detection of torsion bar twist angle due to torque is one of the purposes of micro-angle detection. Fig. 8 shows the structure of a sensor used for this purpose. Specifically, a multipole magnet is placed at one end of a torsion bar, and yokes are placed at the other end, making it possible to detect relative twist angles during rotational motion. However, in the case of the structure as is in Fig. 6, rotation occurs up to the Hall IC, thereby causing hindrance.. To solve this problem, magnetic flux collecting rings are arranged in the vicinity of the outer peripheries of the yokes in such a way that the magnetic flux in the yokes is taken out into the magnetic flux collecting rings so that the magnetic flux between the upper and lower flux collecting rings will be detected by the Hall IC. With this arrangement, the magnetic flux collecting rings and the Hall IC can be held fixed without being rotated, thus making it easy to take out wiring. Furthermore, if multiple Hall ICs are installed, it is possible to easily enhance redundancy.

Fig. 9 shows the output characteristics of this sensor. The maximum angle detection range of this sensor was set at 3 degrees. As shown in the figure, it was possible to obtain output characteristics with satisfactory linearity up to a twist angle of 3 degrees.

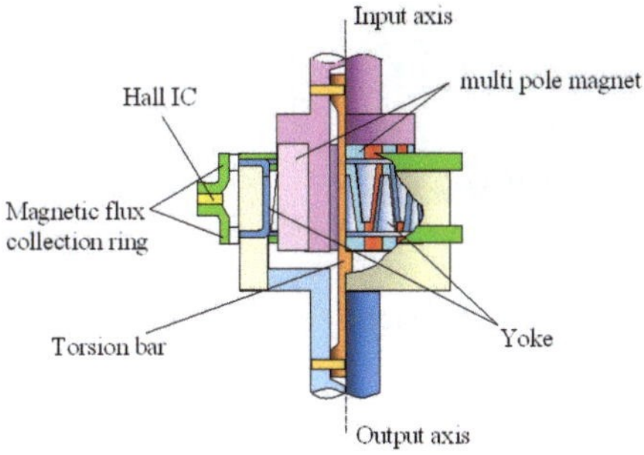

Fig. 8. Structure of micro-angle sensor

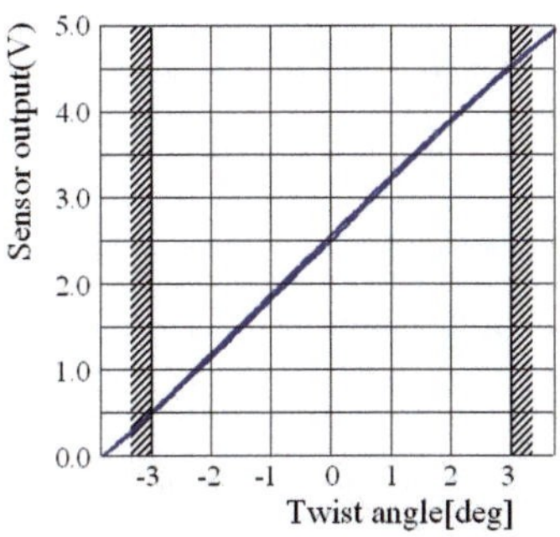

Fig. 9. Output characteristics of this sensor

5 Magnetic Field Simulation Analysis

In establishing the specifications, including the dimensions, of the above-mentioned sensor, we utilized magnetic field simulations to study the shapes of relevant components. In order to improve accuracy, it is desirable that magnetic flux density variations be large with respect to rotation angles. The shape of the yoke tabs also has a great effect on this matter. As an example, Fig. 10 shows the results of an analysis of the effect of varying the short side of trapezoidal tabs. Furthermore, the placement of this micro-rotation angle sensor is important when consideration is given to the fact that magnetic flux density is detected by the Hall IC in this micro-rotation angle sensor. Fig. 11 shows the results of the simulation used in this study.

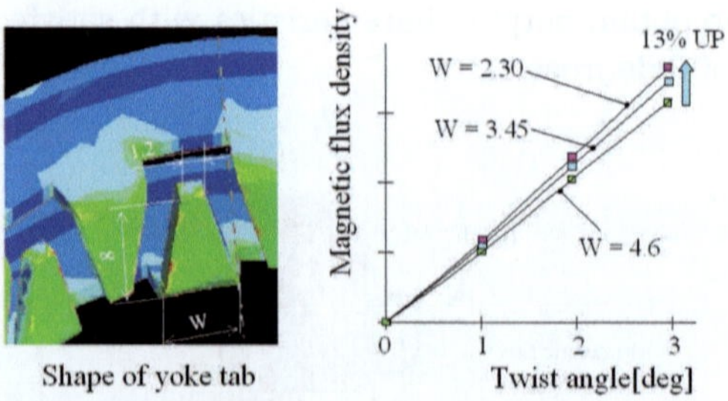

Fig. 10. Study of tab dimensions

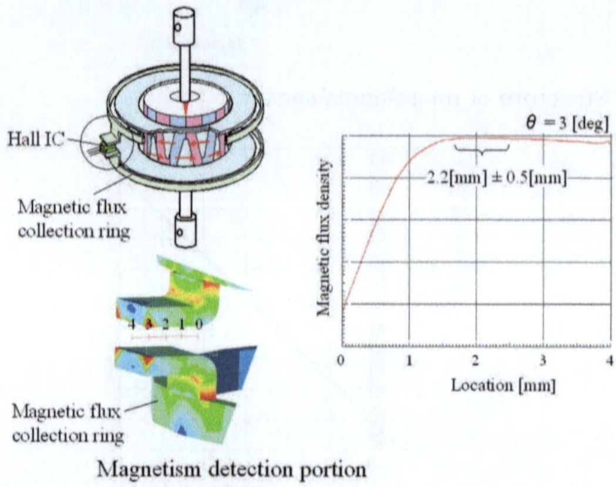

Fig. 11. Study of Hall IC placement

6 Conclusions

To design a micro-rotation angle sensor with excellent vehicle mountability, the authors developed a small sensor which uses a new magnetic circuit and a Hall IC and which also have excellent environmental resistance. This sensor is featured by a structure consisting of a multipole magnet, yokes with tab portions that concentrate the magnetic flux from the magnet, and magnetic flux collecting rings. In this way it was possible to achieve both high accuracy and high sensitivity with the magnetic type sensor.

References

[1] "Automotive Sensors & Sensor Interfaces" Dirk Hammerschmidt, Patrick Leteinturier, 2004 SAE World Congress, March 8-11,2004, Detroit,USA

[2] "An Evolution of Vehicle Control Systems Supported by Sensing Technologies", Akira Asaoka, JSAE Vol.61 2007, pp.10-15

[3] "A Hall Effect Rotary Position Sensor", Robert Bicking, George Wu, Joe Murdock, Don Hoy, Rusty Johnson, SAE 910270,1991

Toshikazu Ina , Kenji Takeda
14 Iwaya Nishio-city Aichi-Pref.
ZIP 445-0012
Japan
Toshikazu_ina@soken1.denso.co.JP
Kenji_x_takeda@soken1.denso.co.jp

Akihiko Sawada, Shigetoshi Hukaya
Denso Corporation
EPS Engineering Center
1-1 Showa Kariya Aichi-Pref.
ZIP 448-8661
Japan
Akihiko_sawada@denso.co.JP
Shigetoshi_hukaya@denso.co.jp

Keywords: rotation angle, hall IC, magnetic sensor, micro rotation angle

Scaled Test Bed for Automotive Experiments: Evaluation of Single Accelerometer Electronic Stability Control

D. I. Katzourakis, ECE, Technical University of Crete
A. I. Katzourakis, Foundation for Research and Technology

Abstract

Evaluation and testing of electronic stability adaptive control systems in automotive vehicle, in the real environment confronts with cost and safety concerns leading to an overdue prototyping of the actual system. This paper addressees a simple Electronic Stability Control Algorithm, evaluated, tested and post-process simulated in a 1:5 scaled model car especially designed for rapid system prototyping. The custom-developed model car acting as experimental platform is fully equipped with sensors, actuators, a controller to collect data and a Linux based computer system to process data. By using a scaled model car we introduce realistic simulation dynamics and disturbances. The reference model for stabilization is based upon the dynamics of the Bicycle Model. The Stability Control System issues commands to the individual car brakes in order to reduce the error between the actual and desired response of the car. The experimental results are enlightening regarding the algorithm's applicability and the vehicle's dynamics.

1 Introduction

Electronic Stability Control (ESC) is a closed loop computer based system which helps the driver to maintain control of the vehicle and prevent skidding under highly demanding situations by applying individual wheel braking and/or readjusting the engine-delivered torque. The driver can be modelled as a high gain system whose reactions are cursory and boorish and might worsen a situation of instability. Even an experienced driver in a panic situation might try to counteract the effect of oversteer (or understeer) in a rear wheel drive vehicle by applying the brakes, an action that will increase the violence of the effect. The loss of handling in such a scenario is probably to result to a fatal accident. Several studies since the wide mass introduction of ESC, in the year 1998, have showed the system's effectiveness [1]. At least 40% of fatal accidents are triggered by skidding and the global installation of ESC could reduce skidding accidents up to 80% [2]. The undisputed benefits from ESC led the European Union to launch a campaign called "Choose ESC" at the Bridgestone European Testing Ground, on 8th May 2007. The aim of this campaign was to spark people's awareness towards ESC and promote the active safety market in automotive industry with the installation of ESC, in all vehicles in European ground. It is prominent that in year 2007, in Germany, Denmark, Austria and Italy, ESC system is standard fit for almost all brands. The goal of the "Choose ESC" campaign was to halve the road fatalities by 2010.

Fig. 1. Scaled model car

Extensive research towards stability and yaw control has been conducted by several authors and different approaches have been proposed [4,5,8] with BOSCH GmbH being the pioneer and thereinafter leader in ESC. The stabilization of the vehicle is accomplished by individual wheel braking [7], active steering [5, 6] and hybrid methods [9] combing the precedents along with dynamic engine's torque distribution. Challenging position on modelling about stability and desired path tracking on rally driving techniques has also been addressed by [10] and [11], where maneuvers commonly used for high speed cornering, like pendulum turn and trail braking have been analyzed. Similar to previous systems should be adapted by automotive manufacturers for their fast fleet cars as an optional cachet in order to increase vehicle's fun to drive side.

ESC course along with fondness towards automobiles and high speed driving, challenged us to develop a 1:5 scaled model car to be used as test bed for experiments in vehicle dynamics and control (Fig. 1). Similar work has been done by [3] and [13]. The rest of the paper is presented in four sections: System's Implementation, Vehicle Dynamics and Stabilization Algorithm, Real Environment Evaluation, Conclusions and Future Work.

2 System's Implementation

The test bed is based on a XRC 1:5 scale remote control car. The model is a 2 rear wheel drive car, with a single cylinder 23cc two stroke air-cooled engine. It has a centrifugal clutch for transmission, a single disk brake for the rear axle and two independent disk brakes for each of the front wheels. It has a standard 2 channel FM radio with one servo for steering and another for throttle-brake with 18 kg-cm and 5.5 kg-cm torsion correspondingly with response time of 0.48 s/60°. In order to meet our experimental standards, the platform had to be completely self-contained from equipment or machinery off the vehicle, had to be inexpensive, independent from special hardware and commercial software and correspond the terms of GNU general public licence. The model is equipped with all the necessary sensors, actuators and computing power for data fusion, dynamics modelling and control. The main processing unit of the system is a mini ITX VIA Computer running Linux. The data logging and actuation control is implemented in an ATMEL 8 bit microcontroller which communicates through the serial port with the computer. The majority of the software for the microcontroller and the computer is written in C/C+ + and all the necessary hardware is built from scratch. The system also has a wireless LAN for remote access through a laptop with IEEE 802.11.

2.1 Hardware

As mentioned before, our goal was to build a low cost experimental platform with the potentiality to alter and publish software and results, without any special licence. We concluded that a mini ITX (form factor 17 cm*17 cm) PC with a free operating system would be appropriate. Hence the system is based on VIA Jetway single board computer (SBC), with onboard VIA C7 1.5 GHz nanoBGA2 Processor, an integrated graphics card, 400 MHz FSB, 512 MB DDR2 400 MHz RAM, 1 PCI slot, ATA 100/133 & 2 SATA ports support, IEEE 1394 firewire, 8 USB 2.0 ports & 2 COM ports and VIA 10/100 Ethernet, running an Ubuntu 6.10 Linux distribution under the GNU general public licence. There has also been installed a wireless g Linksys WRt54GC compact router connected to the

SBC via the Ethernet port. The SBC is powered through 12 V 80 W pico power supply and the power source is 12 V, 4 Ah NiMh battery pack. Data fusion and actuation control is conducted by a versatile custom designed 2-layer PCB. The PCB has an attached AVR ATMega32 RISC microcontroller connected to the serial port of the SBC; with 32 Kb program flash and 2 Kb RAM. USB communication was prohibitive, because it would consume the majority of computing power of the microcontroller running at 16 MHz. ATMega32 doesn't have an embedded USB core, so the communication would have to be incorporated in software. The speed of data flow is 115.2 Kbs which is more than adequate for our purpose, since the bottleneck of the system is the actuation and not the computing power. The microcontroller has been exploited to the limits, since it manages to control 6 actuators (servo) and log data from 17 sensor inputs, in a sum of 32 I/O ports with an astounding real time precision.

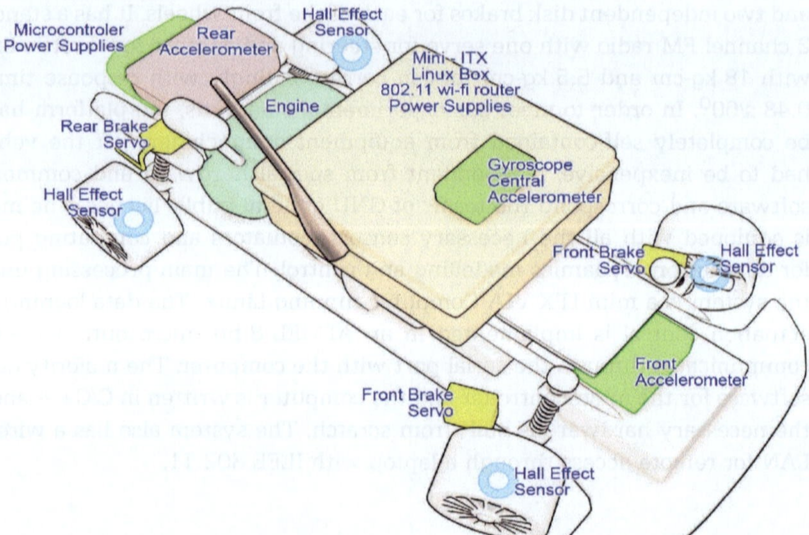

Fig. 2. Hardware Placement.

2.2 Sensors and Actuators

Sensors:

▶ Lateral and longitudinal acceleration in the front and rear axle of the vehicle is evaluated through two ADXL311 accelerometers from Analog Devices. ADXL311 is a low cost $\pm 2g$ accelerometer with analog output, proportional to the measured acceleration. The output of this sensor (X and Y axis measurement) is fed to the microcontroller through an analog to digital converter (ADC) I/O line. Lateral and longitudinal acceleration for the center of gravity of the car, is evaluated via one ADXL213 $\pm 1.2g$ accelerometer from Analog Devices. The output of this sensor is pulse width modulated and the acceleration can be realized from the duty cycle of the signal. The signal (from both X and Y axis) is fed to a XOR gate, and the output of the XOR gate is fed to the first of three external interrupt lines of the microcontroller. The signal is also driven to two I/O lines. The external interrupt is programmed to be triggered by any logical change. Hence, any logical change from the output of the ADXL213 will signal an interrupt, allowing the microcontroller to determine the duty cycle from X or Y axes with a single interrupt.

- Steering angle estimation is conducted via variable resistance. An articulated mechanism moves the shaft of a potentiometer, with motion derived from the steering axon. The output of the potentiometer is driven to an ADC I/O line. Because the voltage generated is not proportional of the actual steering angle [14], we have sampled enough voltage values with respect to the actual angle of each wheel. With the built in function of MATLAB, polyfit (uses LMS to fit values to a polynomial), we have derived to a third degree polynomial of angle with respect to voltage. The actual and estimated angle is quite close.
- Wheel angular velocity is computed with four SS433 Hall effect sensors from Honeywell. When exposed to the appropriate magnetic flux, SS443 sinks its output. 8 reed relay magnets have been attached at each rim. The angular velocity is updated every 100 msec, delivering 2*pi/(8*100 ms) rad/sec or 0.4625 m/s (longitudinal speed with 11.7 cm wheel diameter) resolution for each wheel. A timer counter interrupt measures the logical transitions for every sensor for a certain amount of time. The same scheme like ADXL213's, using one I/O external interrupt and a XOR gate has been used here too.
- Yaw rate estimation is conducted through an ADXRS300 gyroscope from Analog Devices. It is a $\pm 300^{o}$/s yaw rate gyro with signal conditioning. The output from the gyro is analog, proportional to the yaw rate, and has 5 mV/^{o}s sensitivity. The output is fed directly to an ADC line for estimation.
- Driver's Command from the transmitter are delivered to the receiver and sampled from the microcontroller. The output of receiver is a pulse width modulated 50 Hz signal. It is sampled in a satisfactory manner, every 32 μs.

Actuators:

- Standard model servo mechanisms are the actuators for the platform. There are two servos attached to the nave of each of the left and right wheel correspondingly and another one for the real axle brake; exclusive of the two factory installed servos. The greatest drawback of the design is the usage of standard servos. A real automobile, with ABS installed has hydraulic valves as actuators and the reaction time is ten times less than an electro mechanical servo. The system can also control the servos for throttle, brake, and steering.

2.3 Software

Single Board Computer:

The operating system installed is Linux Ubuntu 6.10. There have been installed an SSH server, the GNU GCC compiler, an FTP server and an Apache webserver. Everything was set to meet the necessary requirements for the robust function of the platform. A daemon server was developed for data acquisition from the computer's serial port. The daemon is written in C/C+ + and besides the data collection calls the routine for the stabilization of the vehicle, also written in C/C+ +. A Java Server/Client program for graphical user interface through a common browser is being developed at the moment.

Microcontroller:

The firmware for the microcontroller was developed in C language, with WinAVR [15] a suite of executable, open source software development tools for the Atmel AVR series of RISC microcontrollers. WinAVR includes the GNU GCC compiler for C and C+ +. The software in total is free and available in WWW.

3 Vehicle Dynamics and Stabilization Algorithm

3.1 Vehicle Dynamics

A single track model is used for our implementation that is derived from the mathematical model considered by [14] for steady state cornering. The model takes into account tractive and inertial around the yaw axis forces, and neglects roll and pitch motion. Nevertheless, the complete dynamics of a real vehicle are highly non-linear and difficult to control. Interesting approaches have been proposed for non-linear systems with measurable state by [16], where an unknown non-linear system is controlled, with the usage of recurrent high order neural networks. The geometry of our single track model is shown in Fig. 3. The dynamics equations are given by 3.1, 3.2, 3.3 and 3.4:

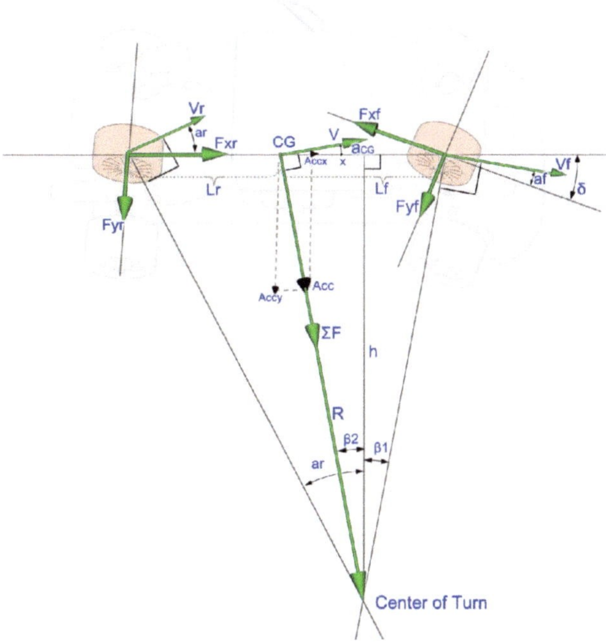

Fig. 3. Vehicle Model

Applying Newton's Second Law in the lateral direction of the vehicle, we can derive the necessary dynamics equations. For a vehicle travelling with a speed of V, the sum of the lateral forces (ΣF_L) originating from the tires acting on the vehicle are equal to the centripetal force.

$$\sum F_L = F_{Lf} + F_{Lr} = M\frac{V^2}{R} \tag{3.1}$$

Where M the mass of the vehicle, R: radius of turn, F_{Lr} and F_{Lf} the sum of the lateral forces (vertical from each wheel's direction of travel) acting on the rear and front axle correspondingly. Applying Newton's second law around the Center of Gravity (CG), if we consider the vehicle to be in an equilibrium moment:

$$F_{Lf} \cdot l_f - F_{Lr} \cdot l_r = 0 \tag{3.2}$$

Where l_f and l_r are the distance of the front and rear axle from the CG correspondingly. If we substitute equation 3.2, back to 3.1:

$$M \cdot \frac{V^2}{R} = F_{Lr} \cdot (\frac{l_r}{l_f} + 1) = F_{Lr} \cdot (\frac{l_r + l_f}{l_f}) = F_{Lr} \cdot (\frac{L}{l_f}) \tag{3.3}$$

$$\left.\begin{aligned} F_{Lr} &= M \cdot (\frac{l_f}{L}) \cdot \frac{V^2}{R} \\ F_{Lf} &= M \cdot (\frac{l_r}{L}) \cdot \frac{V^2}{R} \end{aligned}\right\} \tag{3.4}$$

If we look at Fig. 3, we conclude:

$$F_{Lf} = F_{yf} \cdot \cos(a_f + \delta) + F_{xf} \cdot \sin(a_f + \delta) = M \cdot (\frac{l_r}{L}) \cdot \frac{V^2}{R} \tag{3.5}$$

$$F_{Lr} = F_{yr} \cdot \cos(a_r) + F_{xr} \cdot \sin(a_r) = M \cdot (\frac{l_f}{L}) \cdot \frac{V^2}{R} \tag{3.6}$$

Key parameters and symbols defined in Tab. 1.

Symbol	*Variables and parameters for the single track model*
M	*Mass of the Vehicle*
V	*Forward speed*
l_f, l_r	*Distance of front, rear axle from the center of gravity*
L	*Wheelbase ($L = l_f + l_r$)*
F_{Lf}, F_{Lr}	*Sum of lateral forces for front and rear axle.*
F_{yf}, F_{yr}	*Cornering forces: front, rear axle*
F_{xf}, F_{xr}	*Tractive forces: front, rear axle*
a_f, a_r	*Slip angles: front, rear axle*
δ	*Steering Angle*
R	*Radius of Turn*

Tab. 1. Key parameters and symbols

Cornering forces F_{yf}, F_{yr}, are linear w.r.t. slip angle at low slip angles [14]. The relationship between forces and slip angle is:

$$\left.\begin{aligned} F_{yr} &= C_{ar}\cdot a_r \\ F_{yf} &= C_{af}\cdot a_f \end{aligned}\right\} \tag{3.7}$$

For angles less than 20°, the error for a first order approximation for sine and cosine ($\sin\alpha \approx \alpha$ and $\cos\alpha \approx 1$), is less than 6.5%. From Fig. 3, assuming small angles, we can derive:

$$\left.\begin{aligned} \tan a_r &= \frac{l_r + x}{h} \xrightarrow{\;a_r\text{ :small}\;} a_r = \frac{l_r + x}{h} \\ \tan(\beta_1) &= \frac{l_f - x}{h} \xrightarrow[\beta_1=\delta-a_f]{\;a_f\text{ :small}\;} \delta - a_f = \frac{l_f - x}{h} \\ \cos(\beta_2) &= \frac{h}{R} \xrightarrow[\beta_2=a_r-\delta+a_f\approx 0]{\;\beta_2\text{ :small}\;} \frac{h}{R} = 1 \end{aligned}\right\} \tag{3.8}$$

Where x is distance from CG to the projection of Center of Turn on the longitudinal axis of the vehicle. Adding the first two and substituting the 3rd, yields:

$$a_r + \delta - a_f = \frac{l_r + l_f}{h} \xrightarrow{\;h=R\;} \delta = \frac{\mathrm{L}}{\mathrm{R}} + a_f - a_r \tag{3.9}$$

If we substitute equations 3.7 into 3.5 and 3.6, we get:

$$C_{af}\cdot a_f\cdot\cos(a_f+\delta) + F_{xf}\cdot\sin(a_f+\delta) = M\cdot(\frac{l_r}{L})\cdot\frac{V^2}{R} \tag{3.10}$$

$$C_{ar}\cdot a_r\cdot\cos(a_r) + F_{xr}\cdot\sin(a_r) = M\cdot(\frac{l_f}{L})\cdot\frac{V^2}{R} \tag{3.11}$$

Again, assuming small angles:

$$\cos(a_f+\delta) = \cos a_f\cos\delta - \sin a_f\sin\delta \approx 1 - a_f\cdot\delta$$
$$\sin(a_f+\delta) = \cos a_f\sin\delta + \sin a_f\cos\delta \approx \delta + a_f$$

Using the above relationships into 3.10 and 3.11:

$$C_{af}\cdot a_f\cdot(1-a_f\cdot\delta)+F_{xf}\cdot(a_f+\delta)=M\cdot(\frac{l_r}{L})\cdot\frac{V^2}{R}\rightarrow$$

$$C_{af}\cdot a_f-C_{af}\cdot a_f^{\,2}\cdot\delta+F_{xf}\cdot a_f+F_{xf}\cdot\delta=M\cdot(\frac{l_r}{L})\cdot\frac{V^2}{R}$$

$$\xrightarrow[if\ a_f,\delta=20^o\rightarrow a_f^{\,2}\cdot\delta=0.04]{a_f^{\,2}\cdot\delta\approx 0} \tag{3.12}$$

$$C_{af}\cdot a_f+F_{xf}\cdot a_f+F_{xf}\cdot\delta=M\cdot(\frac{l_r}{L})\cdot\frac{V^2}{R}$$

And

$$C_{ar}\cdot a_r\cdot 1+F_{xr}\cdot a_r=M\cdot(\frac{l_f}{L})\cdot\frac{V^2}{R} \tag{3.13}$$

Solving 3.12 and 3.13 w.r.t. a_f and a_r yields:

$$a_f=\frac{M\cdot l_r\cdot V^2}{C_{af}\cdot(1+\frac{F_{xf}}{C_{af}})\cdot R\cdot L}-\frac{F_{xf}\cdot\delta}{C_{af}\cdot(1+\frac{F_{xf}}{C_{af}})} \tag{3.14}$$

$$a_r=\frac{M\cdot l_f\cdot V^2}{C_{ar}\cdot(1+\frac{F_{xr}}{C_{ar}})\cdot R\cdot L} \tag{3.15}$$

If we substitute equations 3.14, 3.15 into 3.9, and solve w.r.t. R we get:

$$\frac{\delta\cdot\left(1+2\cdot\frac{F_{xf}}{C_{af}}\right)}{(1+\frac{F_{xf}}{C_{af}})}=\frac{1}{\mathrm{R}}\cdot\left(L+\frac{M\cdot l_r\cdot V^2}{C_{af}\cdot(1+\frac{F_{xf}}{C_{af}})\cdot L}-\frac{M\cdot l_f\cdot V^2}{C_{ar}\cdot(1+\frac{F_{xr}}{C_{ar}})\cdot L}\right)\Leftrightarrow$$

$$\left.\begin{array}{l} R=A\cdot\left(L+\left(\frac{V^2\cdot M}{L}\right)\cdot(B-C)\right) \\ A=\frac{1}{\delta}\cdot\frac{\left(1+\frac{F_{xf}}{C_{af}}\right)}{\left(1+2\cdot\frac{F_{xf}}{C_{af}}\right)}\quad B=\left(\frac{l_r}{C_{af}\cdot(1+\frac{F_{xf}}{C_{af}})}\right)\quad C=\left(\frac{l_f}{C_{ar}\cdot(1+\frac{F_{xr}}{C_{ar}})\cdot}\right) \end{array}\right\} \tag{3.16}$$

If we consider a steady steering angle δ, we can analyze the behaviour of the radius in a turn for each term (A, B and C) of the above equation.

3.2 Oversteer and Understeer Effects

When a vehicle oversteers it tends to narrow the radius of turn. On the other hand, when a vehicle understeers, tends to widen the radius (Fig 4). Applying the right inputs when the effect of oversteer or understeer is detected, the ESC system can counteract the undesired effects within the physical limits of the system. With a closer look at equation 3.16 we can determine which would be the appropriate inputs for each case.

Oversteer: In order to counteract oversteer; we have to increase the radius of turn:

- Reduction on the magnitude of a positive F_{xf}, or better a negative F_{xf}, increases both A and B terms, consequently R. This is translated as releasing throttle on a FWD vehicle or applying brakes at the front axle.
- Increase on the magnitude of a positive F_{xr} reduces C terms, therefore increases R. This accounts for applying throttle on RWD vehicle.

Understeer: In order to counteract understeer, we have to decrease radius of turn:

- Increase on the magnitude of a positive F_{xf} reduces both the A and B terms, therefore R. This is translated as applying throttle on FWD vehicle.
- Reduction on the magnitude of a positive F_{xr}, or better a negative F_{xf}, results to an increased C term, therefore R. This accounts for releasing

throttle on a RWD vehicle, or applying brakes at the rear axle (tail-brake).

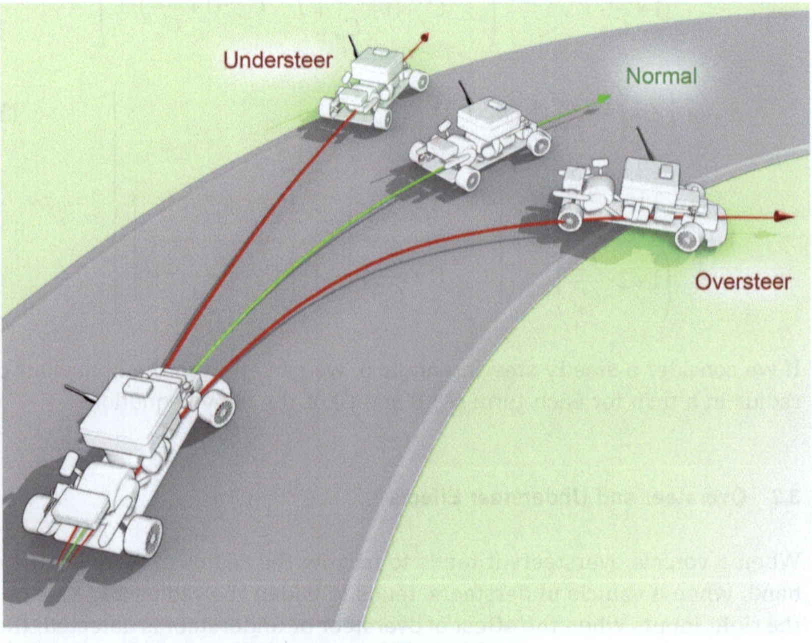

Fig. 4. Oversteer, Understeer and Neutral Steer

3.3. Stabilization Algorithm

The stabilization scheme has been designed in order to be simple to implement, easy to adjust and would be cost effective on a real vehicle, considered an ABS system already installed. It could be implemented with a single dual axis accelerometer and a simple microcontroller with an embedded CAN Bus core, in order to be able to cooperate with the ABS controller and the ECU of the vehicle.

The radius of turn for a vehicle travelling with low speed, with absence of oversteer or understeer, assuming small angles will be equal to $R_m = L/\delta$ [14]. This is the inverse of Ackerman steering angle, for a desired radius of turn. We can use the inverse of Ackerman's angle as a reference model for driver's desired radius of turn. The actual turning angle can be determined from a dual axis accelerometer, installed at the CG of the vehicle. The accelerometer can measure lateral acceleration towards the center of turn. The relationship is:

$$Acc = \frac{V^2}{R} \tag{3.17}$$

Through the accelerometer, we can determine the magnitude of slip angle for CG. The direction of travel for the vehicle at CG is vertical to the vector of lateral acceleration. The dual axis accelerometer is oriented so that can measure both lateral (Acc_y) and longitudinal (Acc_x) acceleration of the vehicle. Therefore, slip angle a_{CG} at CG (Fig.3) can be determined by:

$$a_{CG} = \tan^{-1}\left(\frac{Acc_x}{Acc_y}\right) \tag{3.18}$$

Since the only available speed for measurement is towards the vehicle's longitudinal axis (V_X), we can compute

$$V \text{ via: } V = V_X \cdot \cos(a_{CG}) \tag{3.19}$$

Thus, through equations 3.17, 3.18 and 3.19 we can determine the radius of turn relatively well.

The stabilization algorithm is built for a RWD vehicle. The actuation is performed on the brakes of front or rear axle. A sensitivity parameter S has been implemented in the algorithm for an adjusted desired understeer or oversteer behaviour of the vehicle.

4 Real Environment Evaluation

The stabilization algorithm proposed, although it looks promising, did not provide the expected results when evaluated on the test bed. Real environment simulation showed that the use of a single accelerometer for electronic stability control is inhibitory for a scaled vehicle. The limited functionality of the stabilization algorithm lies in the physical operation of the accelerometer. Vibration from the engine (two stroke single cylinder engine) and the anomalies from the ground in combination with small size of the vehicle corroded the measurements from the accelerometer. Although the accelerometer ADXL213 has provisions for bandlimiting the measurements and can achieve low pass filtering for antialising and noise reduction, it was physically constrained by the harsh environment of operation. Fig. 6 shows static vibration sampled from the ADXL213 w.r.t. time. The bandwidth for the accelerometer was set at 50 Hz, much lower than the frequency operation of the engine in the lower speed, which is about 800 rpm, for a typical two stroke engine. The selection of the bandwidth for the accelerometer was a trade off between necessary measurements and noise reduction. 50 Hz was the best choice for our implementation. The outcome from the miss measurements was a miss calculated actual radius of turn. In most cases, the algorithm detected understeering, thus applied brakes on the rear axle, provoking the vehicle to oversteer.

1) Evaluate Driver's Desired Radius of Turn

$$R_m = \frac{L}{\delta}, \quad \delta\text{: steering input in radians}$$

2) Estimate Actual Radius of Turn

$$R = \frac{V^2}{Acc} \quad \left\{ \begin{array}{l} V = V_X \cdot \cos(a_{CG}) \\ a_{CG} = \tan^{-1}\left(Acc_x / Acc_y\right) \end{array} \right\}$$

3) If $R < S \cdot R_m$ *

Apply Instantaneous Brakes on Front Axle

Else if $R > (1/S) \cdot R_m$

Apply Instantaneous Brakes on Rear Axle

Else

Do nothing

4) Repeat Forever

* S: Sensitivity $0 < S < 1$

Fig. 5. The ESC Algorithm

Another drawback of the algorithm that was clarified from experimental data is that the reference radius used in the algorithm is valid only for very slow speed turning. Measurements from the rest of the sensors revealed that the necessary centripetal force for the vehicle to follow the desired radius of turn is difficult to be produced in real environment. From the data, we can also derive the maximum yaw rate, before the vehicle loses control, with respect to individual wheel speed and steering angle. Loss of control can be determined from the behaviour of vehicle's yaw rate. That means if we have a constant steer angle and small variations at the speed of front wheels (back wheels might be spinning, thus are bad candidates for remarks) and we experience great variations on yaw rate if the vehicle oversteers. On the other hand, if we have an increase at the speed of vehicle's front wheels and/or increase of steer angle, and experience little or decrease in magnitude of the yaw rate, the vehicle understeers. Fig. 7 and Fig. 8 present yaw rate development in time, in perspective with individual wheel speed and steering angle.

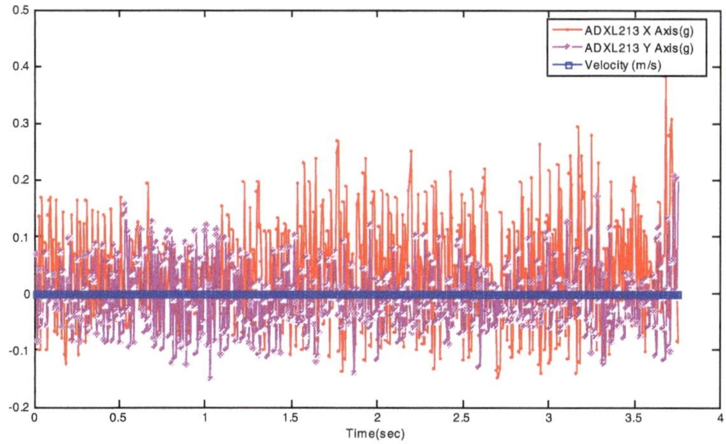

Fig. 6. Static Acceleration from Vibration

Fig. 7 shows a loss of handling situation. At time around 0 sec the driver steers the vehicle to the left. The driver realizes that the car oversteers, and tries to counteract by turning the wheels at the opposite direction (right) and applying throttle around 0.5 sec. The rear right wheel starts spinning and after almost a second of drifting the vehicle stabilizes at 2.5 sec. Fig. 8 shows a different loss of handling situation. From time equals 0 sec until 1.4 sec the vehicle oversteers very intensely. Notice at time equals 1 sec that the rear left wheel has the lowest speed of all, since it is the inner wheel w.r.t. a left wise turn; thus has the smallest trajectory. At 1.2 sec the driver applies throttle inducing the rear left wheel to start spinning. Application of throttle at the rear axle and the turning at the opposite direction (right), produces moments of understeer at time = 1.5 sec. At around 1.4 sec, the electronic stability system, applies brakes at the front axle increasing understeer.

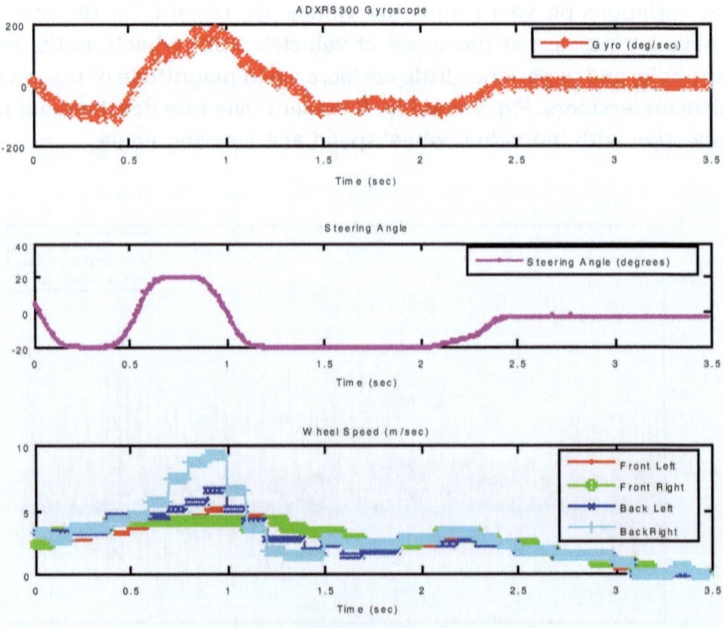

Fig. 7. Yaw Rate Development in Extreme Oversteer

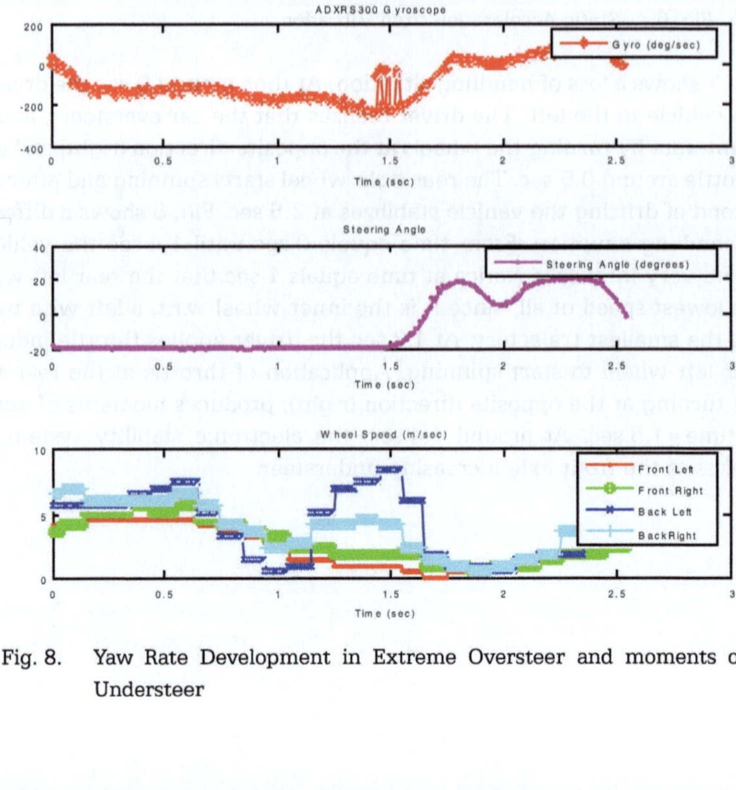

Fig. 8. Yaw Rate Development in Extreme Oversteer and moments of Understeer

5 Conclusions and Future Work

5.1 Conclusion

A scaled test bed was presented along with the evaluation of an electronic stability control scheme. Although the applicability of the proposed ESC system is limited, many useful results came up from the trial run. First of all the use of gyroscope is one – way solution for an ESC system on a scaled test bed. All three accelerometers embedded in the scaled car, reported similar to ADXL213 corruption at measurements. There are several factors that can undermine the credibility of the samples. Vibration from the engine, shock from a bump on the road, inclination of the ground etc. The effects from the precedent would be reduced in a full scale vehicle, compared to a scaled car, but they wouldn't obliterate. The collected experimental data provide also the opportunity for a better understanding of understeer and oversteer effects. Thereupon, we might be able to derive more robust ESC schemes and heuristic algorithms. Through examination of velocity, steering angle and yaw rate we might be able to determine the maximum yaw rate a vehicle can experience before it loses handling. Thus, it is possible to construct an off – line reference function for yaw rate and discern when the vehicle experiences an undesired effect.

If we prove an explicit connection between the test bed and a real vehicle, the results and algorithms that would derive from the scaled car would be applicable to a real vehicle too. Therefore, the use of scaled test beds in automotive engineering could accelerate the prototyping process and diminish the cost of development.

5.2 Future Work

- Limit the bandwidth for the accelerometers to 10 Hz. Implement more magnets in the wheels, in order to decrease sampling period for wheel speed (100 ms at the time) and increase together accuracy in measurements. Add faster and more powerful actuators (higher torque) on the model.
- Manifest the connection between experiments in the test bed and a real vehicle.
- Use car dynamics based on double track model and investigate vehicle's behaviour under high speed driving techniques [10, 11].
- Design appropriate yaw rate reference function and develop an adaptive algorithm for stabilization.

Implementation of similar sensory provision on a real vehicle is desired.

6 Acknowledgements

The authors are grateful to Professor Manolis Christodoulou at TUC, Professor Petros Ioannou at the University of Southern California and Graduate student Spiros Ninos at TUC. The project is sponsored by technical University of Crete.

References

[1] European New Car Assessment Programme (EuroNCAP) and European Commission, "Choose ESC", 2007.

[2] National Highway Traffic System Administrator, "Electronic Stability Control System," FMVSS no.126, March 2007.

[3] Paul Yih, "Radio Control Car Model as Vehicle Dynamics Test Bed," Dynamic Design Lab, Stanford University, September 2000.

[4] J. Ackermann, "Robust Car Steering by Yaw Rate Control," Conference Proceedings of the, 29th Conference on Decision and Control, Honolulu, Hawaii December 1990.

[5] J. Ackermann, T. Bünte, D. Odenthal, "Advantage of Active Steering for Vehicle Dynamics Control," German Aerospace Research Establishment, Institute for Robotics and System Dynamics, Oberpfaffenhofen, 1999.

[6] J. Ackermann, T. Bünte "Automatic Car Steering Control Bridges Over the Driver Reaction Time," German Aerospace Research Establishment, Institute for Robotics and System Dynamics, Oberpfaffenhofen, 24 May, 1995.

[7] J. Tjønnas, T. A. Johansen, "Adaptive Optimizing Dynamic Control Allocation Algorithm for Yaw Stabilization of an Automotive Vehicle using Brakes," Med's '06. 14th Mediterranean Conference on Control and Automation, June 2006, pp. 1-6.

[8] H.E.Tseng, B. Ashrafi, D. Madau, T. Allen Brown, D. Recker, "The development of Vehicle Stability at Ford," IEEE/ASME Transactions on Mechatronics, Volume 4, Issue 3, Sept 1999, pp. 223 – 234.

[9] B.A. Guvenc, T. Acarman, L.Guvenc, "Coordination of Steering and Individual Wheel Braking Actuated Vehicle Yaw Stability Control", Intelligent Vehicles Symposium, IEEE Proceedings, June 2003. pp. 288 – 293.

[10] Velenis, E., Tsiotras, P., and Lu, J., "Aggressive Maneuvers on Loose Surfaces: Data Analysis and Input Parameterization," 15th IEEE Mediterranean Control Conference, June 26-29, Athens, Greece.

[11] Velenis, E., Tsiotras, P., and Lu, J., "Modelling Aggressive Maneuvers on Loose Surfaces: The Cases of Trail-Braking and Pendulum-Turn," European Control Conference, Kos, Greece, July 2-5, 2007.

[12] A. T. Van Zanten et al, VDC, the Vehicle Dynamics Control System of Bosch, SAE 950759, 1995.

[13] S. Brennan, A. Alleyne, "Using a Scale Testbed: Controller Design and Evaluation," IEEE Control Systems Magazine, June 2001.

[14] T.D. Gillespie, Fundamental of Vehicle Dynamics, SAE, 1992.

[15] WinAVR, http://winavr.sourceforge.net/ .

[16] G. A. Rovithakis, M. A. Christodoulou, Adaptive Control With Recurrent High-order Neural Networks: Theory and Industrial Applications, Springer 2000.

Diomidis Ioannis Katzourakis
Technical University of Crete
Vernadaki 16B
Chania
Greece
diomidis@systems.tuc.gr
diomkatz@gmail.com

Antonis Ioannis Katzourakis
Foundation for Research and Technology
Sifaka 22
Heraclio
Greece
tonykatz@forth.isc.gr

Keywords: scaled test bed, model car, electronic stability control, sensors, gyroscope, ADXRS300, accelerometer, ADXL213, ADXL311, experiment, vibration, radio control, single board computer, microcontroller, AVR, ATmega32, remote access, hall effect sensor, SS443, single track model, bicycle, oversteer, understeer, yaw rate

A New Micromechanical Pressure Sensor for Automotive Airbag Applications

B. Adam, T. Brandt, R. Henn, S. Reiss, M. Lang, Ch. Ohl
Robert Bosch GmbH

Abstract

Automotive restraint systems today are more and more using pressure sensors in order to early and safely detect side crashes. A reliable, precise and fast measurement of the air pressure is the key precondition for an optimized system performance.

We present a new generation of Bosch pressure sensors called PPS2 (Peripheral Pressure Sensor) that is specifically designed for airbag system application.

The key component of the sensor is a pressure sensor module (SMD187) that integrates a micromechanical membrane, an ASIC with band-pass filter, trimming, and signal processing as well as the digital interface. The micromechanical part is realized using the new pressure sensor technology developed by Bosch named APSM (Advanced Porous Silicon Membrane). APSM purely bases on surface micromechanical process steps and turns out to be very robust monocrystalline Silicon membranes.

The sensor provides a filtered and normalized differential pressure signal $\Delta P/P_0$ to the airbag ECU using the well-established PSI5 standard current interface. The pressure range of the sensor allows to run the system at altitudes up to 4000 m above sea level. The normalized differential pressure signal is transferred within a range of -5 to +15% of the ambient pressure.

PPS2 allows easy application and system setup, offers high system flexibility and provides a reliable, precise and fast pressure signal to the airbag system.

1 Introduction

In 1996, the first airbag pressure sensor for detection of side crashes was introduced. Due to increased performance requirements in regards to side crash protection pressure sensors have become widely utilized. Today, pressure sensors are common practice for fast and robust detection of side crashes in middle and upper class vehicles. Mostly, the performance of airbag systems gets optimized by combining pressure sensors with acceleration sensors. By using systems with suitable combinations of pressure and acceleration sensors the system firetimes and the robustness against misuses can often be improved compared with systems that are purely based on acceleration sensors.

The crash detection bases on measuring the increased pressure inside the door cavity due to the deformation of the door during an impact where the sidedoor gets hit. Pressure signals used for firing decisions often allow a faster and safer distinction of different kinds of crashes. The signals are directly correlated to the deformation of the door and thus to the crashseverity. The sensor is mounted inside the side door cavity. It can be placed either in the dry-room or in the wet-room area. However, the sensor always senses the pressure inside the outer wet-room cavity and thus must always be connected to the wet-room area of the side door.

The second generation pressure sensor PPS2 by Bosch provides a pressure sensor solution specifically designed for usage in airbag systems. The technical performance of the sensor has been further improved and the compliance with the new PSI5 interface standard is given. PSI5 is a well-proven current interface standard providing highly reliable and fast data transmission and enabling easy system setup and high system flexibility at very low cost. Bandpass filter of the pressure signal, signal conditioning and signal processing are integrated into the sensor. The sensor provides a normalized differential pressure signal which is independent of temperature and ambient pressure. By this means, system resources can be saved and crashalgorithms do not need to reflect ambient conditions.

Additionally the pressure sensor PPS2 optionally can transmit the internally available absolute ambient pressure to the electronic control unit (ECU). The absolute pressure can either be used for other applications or control units within the car or it can be used for system diagnosis by comparing absolute pressures of different airbag pressure sensors. This feature allows to improve the system diagnosis without causing additional effort or cost because pressure sensors in airbag systems are always used in pairs.

Technical requirements to the new generation of pressure sensor are:

- the application must be possible in wet- and dry-room door cavity
- the mounting process must allow electronic documentation
- the pressure sensor must allow full system functionality in heights of -100 to 4000 m above sea level, the signal range for normalized differential pressure must consider pressure changes of -5 to 15%
- the current consumption must be reduced to allow compatibility to acceleration sensors
- an option must be implemented to transmit the absolute pressure to the ECU
- the sensor performs internal self-diagnosis and indicates the failure if an error is detected
- an electronic label is sent by the sensor during the initialization phase that allows identification of the pressure sensor in the ECU

Key component and main innovation of the new pressure sensor is the new sensor module SMD187. The sensor module is exclusively developed for usage in airbag systems. It integrates the micromashined pressure sensor element and an ASIC. Besides the sensor module no active components and only few external passive components are needed. The pressure sensor element is fabricated using the new Bosch APSM (Advanced Porous Silicon Membrane) process. The ASIC uses CMOS-technology. ASIC and sensor element are packaged into a mold-premold housing. The 2nd - level package of the sensor has been developed based on well-proven technologies. The package design enables easy and robust mounting of the sensor and provides the required protection of the electronic circuitry while at the same time allowing dynamic pressure changes to reach the pressure sensing element.

2 Basic Aspects of Using Pressure Sensors in Restraint Systems

2.1 Physical Principle and Reason for Using Pressure Sensors

The following considerations apply to the physical principle for using pressure sensors for side-impact detection.

In the case of deformation of an automobile side door, the volume of the door (and thus the pressure in the cavity) changes in a very short period of time. Because the event occurs too quickly for heat to be exchanged with the environment, the pressure change can be calculated using the adiabatic state equation as follows:

$$\Delta V = V_0 \frac{1}{\kappa} \cdot \frac{\Delta P}{P_0} = const \cdot \frac{\Delta P}{P_0} \tag{1}$$

ΔV represents the change in the volume of the cavity, ΔP the increase in pressure, and P_0 the ambient pressure (initial state). The signal $\Delta P/P_0$ is thus proportional to the change in volume and correlates directly to the crash intensity and risk of injury. The entire door is considered as a detection area. It is important to note that the signal amplitude and shape are essentially determined by the impermeability of the door, the geometry of the cavity, and the point of impact on the door in the crash because the percentage change in volume and pressure can be larger or smaller, depending on these conditions, with otherwise constant crash parameters.

Fig. 1 shows sample signal waveforms for different crash conditions. The correlation with crash intensity is clearly recognizable, as well as the fact that a distinction between deployment-relevant crashes and the illustrated no-fire crash can be made in only a few milliseconds. This time scale is compatible with the very high requirements for side impact detection where, due to a minimal crush zone, only a few milliseconds are available for the deployment decision. This is demonstrated by the consideration that a vehicle with a speed of 50 km/h requires only 0.7 ms to travel a distance of 1 cm.

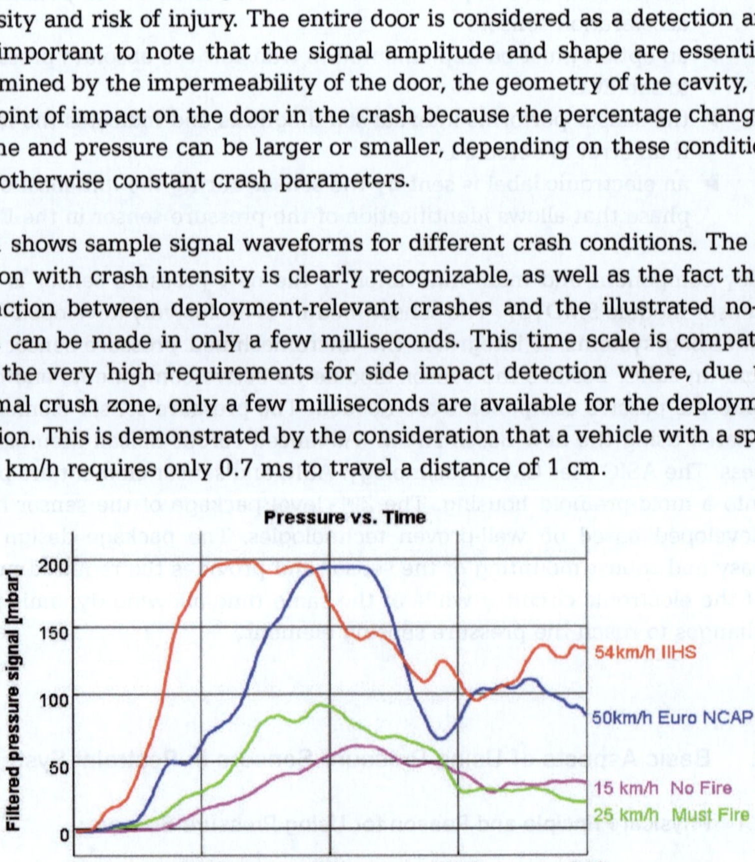

Fig. 1. Pressure signal waveforms during side impacts with various intensities.

A further important property of measuring the pressure in the door volume is the independence of the sensor signal from the exact mounting point of the sensor inside of the door. Fig. 2 shows the simultaneously recorded pressure waveform for 3 different mounting locations inside the door, which illustrates the independence of mounting location. This is a major difference relative to accelerometers, which are typically mounted on vehicle frames or underbodies because this provides reliable, interference-free acceleration data that is maximally decoupled from the contact point of the vehicles involved in a crash. There is another related benefit of using pressure sensors: particularly in side impacts with high points of impact (see Fig. 3), the doors are typically deformed quickly, with a delay in transferring the accelerations to the rest of the body. In such situations with high points of impact, pressure sensors enable clearly better recognition of side impacts than with purely using accelerometers mounted on vehicle underbodies or frames. Corresponding crash requirements result from the large number of relatively high-standing SUV vehicles and are incorporated in crash tests, such as the IIHS Test.

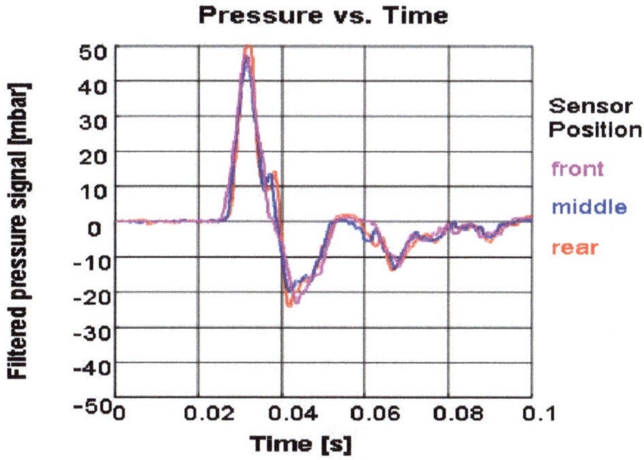

Fig. 2. Measured signal waveforms of pressure sensors mounted at different locations inside a door

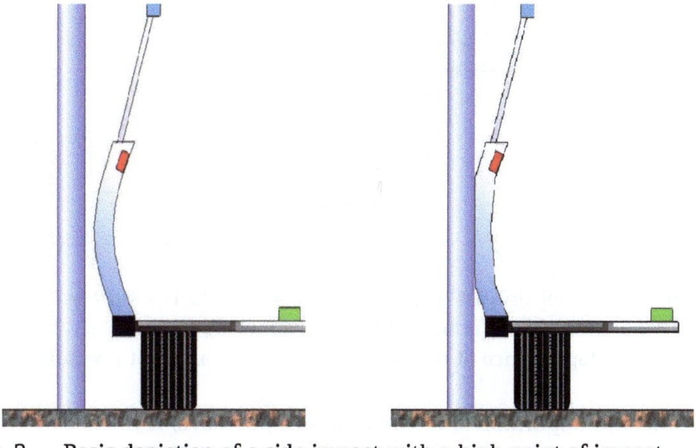

Fig. 3. Basic depiction of a side impact with a high point of impact

In summary, the following conclusions can be drawn with regard to using pressure sensors for side impact detection:

- Pressure sensors support quick, robust deployment decisions
- the physical measurement principle provides a direct link to the risk of injury and the crash intensity
- mounting locations inside the door can be chosen flexibly
- pressure sensors are ideal for detecting side impacts with high impact points

2.2 Functional and Signal Processing Requirements

The functional and signal processing requirement for pressure sensors used in automotive airbag application is that the sensor signal measured during a crash must be independent of the following parameters: ambient pressure (influenced by weather and elevation) and ambient temperature.

If this is not assured, these effects must be taken into account in the algorithms in the control unit, which would lead to increased computational effort and is not desirable.

The following considerations apply to temperature dependence: the pressure profile during a particular crash must not vary as a function of temperature. This means that the sensor must ensure that pressure measurement is independent of temperature. If the measurement principle of the sensor does not assure this, temperature compensation must be provided in the sensor.

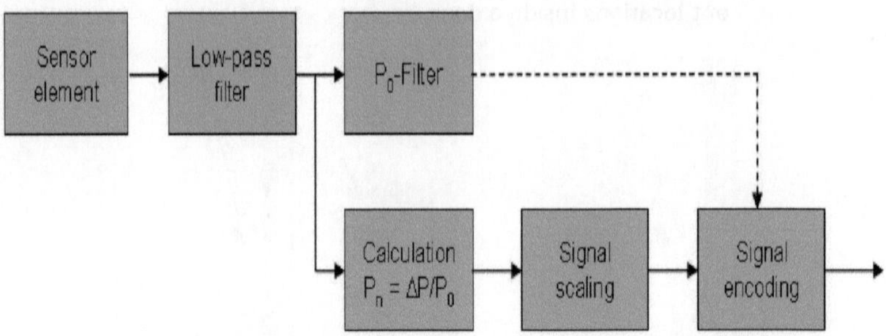

Fig. 4. Signal processing principle of an airbag pressure sensor. A normalized differential pressure signal P_n is necessary to eliminate dependence of the crash signal on the ambient pressure.

Regarding the dependence of the ambient pressure, it must be noted that the pressure profile of a crash varies depending on the ambient pressure. It follows from Eq. 1 that with constant deformation of the vehicle body, the pressure increase in the cavity is proportional to the ambient pressure. Pressure sensors for use with airbags thus have internal signal processing and generate a normalized differential pressure as the useful signal for the algorithm:

$$P_n = S \cdot \frac{\Delta P}{P_0} = S \cdot \frac{P - P_0}{P_0} \tag{2}$$

Here S is a scaling factor that amplifies the signal according to the digitally available signal range of the interface between the sensor and the control unit. This factor is therefore also called the digital sensitivity.

P_0 represents a slowly varying ambient pressure. It is derived from the currently measured pressure P. A low-pass filter adapts P_0 to varying ambient pressures and simultaneously ensures that a constant absolute pressure reference is available during the course of the crash.

In order to filter out high-frequency interfering signals and reduce the noise sensitivity of the sensor sensitivity, pressure sensors and airbag accelerometers both have internal low-pass filters with a cut-off frequency of a few hundred hertz.

The basic elements of the signal processing required in airbag pressure sensors are shown in Fig. 4.

3 Mounting and Environmental Requirements

As described in Section 1, pressure sensors are integrated in airbag systems as an alternative or supplement to peripheral accelerometers. The pressure sensors are mounted inside of the vehicles side doors. The interior of current side doors is usually divided into a wet space and a dry space. The dry space houses devices such as control units and loudspeakers. The wet space houses items such as retracted side windows.

The pressure sensor must be mounted such that the pressure of the outer wet space is detected. In the event of a side impact, this cavity is directly impacted and measuring pressure at this point will deliver the earliest signal.

The sensor and its connection can be located either in the wet or dry space as long as it is ensured that the pressure inlet has free access to the volume of the wet space.

The typical mounting locations are shown in Fig. 5. Depending on the sealing concept of the door, seals may be required on the sensors to ensure that the sensor mounting holes in the sheet metal or partition wall are completely sealed.

The environmental conditions to which the sensor is exposed are derived directly from the installation space. Typical requirements are resistance to immersion, water spray, humidity, humidity variations, salt spray and other media that may be present in the door cavity.

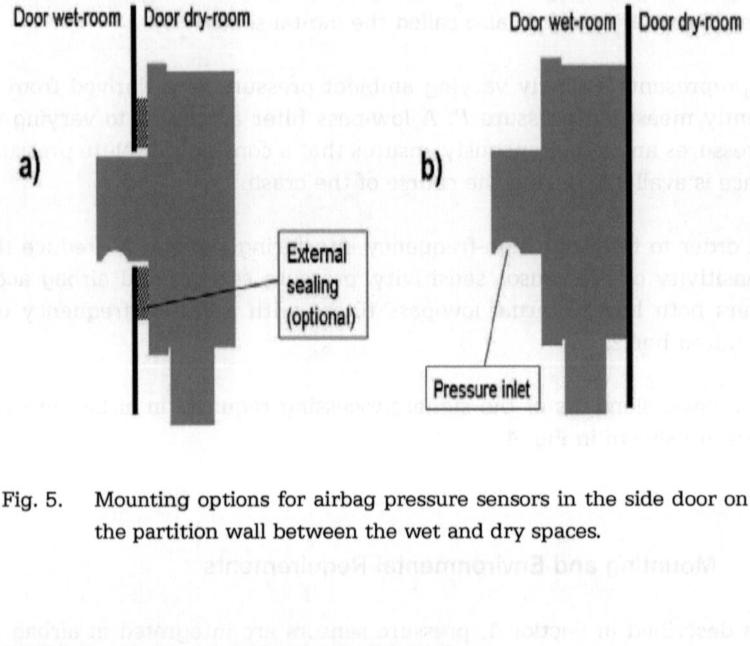

Fig. 5. Mounting options for airbag pressure sensors in the side door on the partition wall between the wet and dry spaces.

4 PPS2 Assembly and Mechanical Interface

A significant improvement in second-generation pressure sensors relative to first-generation sensors is the degree of integration of the electronics.

The micromechanical sensor element and all active electronic components are integrated in a single component: the SMD187 sensor module. Only a few passive external components are necessary in addition to the sensor module. The sensor module and the external circuitry are packaged in an external case called the second-level package. The second-level package provides the mechanical and electrical interface to the restraint system and the car. This is described in more detail below, but first the structural concept of the case and the required processes are explained.

4.1 Structural Concept of the PPS2

The second-level package is essentially based on known an proven processes. A sample is illustrated in Fig. 6, and the sensor structure is shown in schematic form in Fig. 7.

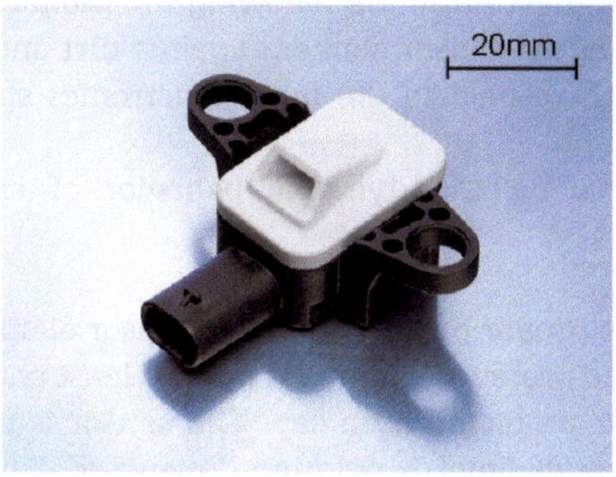

Fig. 6. Sample of a new-generation airbag pressure sensor with cover

The sensor module is mounted on the circuit board with only a few external passive components using reflow soldering. The circuit board is fixed and contacted inside of the external housing by a press fit technology. The press-fit pins are molded into the housing. They provide the contact between circuit board and housing and additionally serve as the connector pins. The press-fit pin provides a mechanical connection between the circuit board and the case as well as the electrical contact. The sensor has a 2-wire interface compliant with the PSI standard, so two pins are sufficient for electrical contact.

The sensor housing is closed by a plastic cover that is attached using laser welding technology. The plastic cover hermetically seals the interior of the sensor.

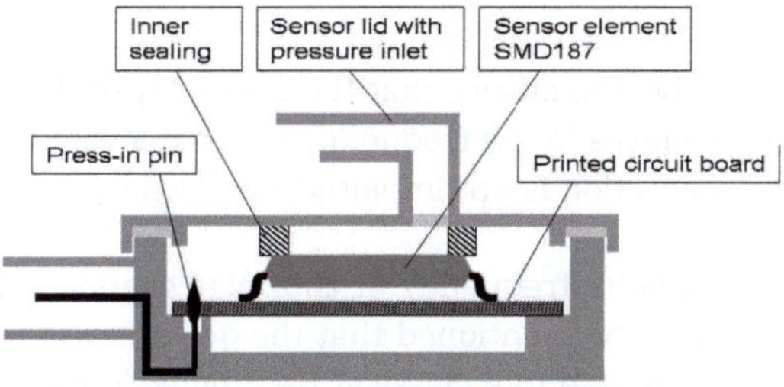

Fig. 7. Schematic diagram of the structure of the airbag pressure sensor: the electronic components are mounted on the circuit board using standard reflow soldering and the circuit board is then pressed into the housing to provide mechanical and electrical connections. The housing is sealed by a laser-welded cover.

A special requirement on the structure of the pressure sensor results from the fact that external pressure must be routed to the pressure-sensitive measuring unit inside the housing. This function is fulfilled by the pressure inlet integrated in the cover.

The pressure inlet must fulfill the following functional requirements:

- Protection of the sensor element against dirt and media
- sealing with respect to the inner electronics space (circuit board and pins)
- no influence on the dynamic behavior of crash-relevant pressure waves

In order to ensure proper protection of the sensor element from the effects of dirt and media, the pressure inlet must have at least one 90° bend. The sensor element is further protected by a cap. The sensor is mounted in the vehicle such that the pressure inlet is pointing downward. The form and size of the inlet is designed so that any dirt or media that penetrates into the inlet will leave due to the effect of gravity.

The impermeability and protection of the interior electronics space is ensured by a seal fitted to the sensor cover around the pressure inlet. When the cover is attached, it seals the inlet channel to the sensor module. This ensures that any media that enters the pressure inlet does not reach the interior electronics space.

In order to ensure adequate dynamic behavior of the sensor with respect to rapid, crash-relevant pressure fluctuations, the pressure inlet is designed to maximize the cross section of the channel and to minimize its length without thereby adversely affecting the previously mentioned protective requirements.

Attention must also be given to ensure that the geometry of the pressure inlet does not allow any resonances in the frequency range relevant to airbag applications so that sensor operation is not impaired by resonance effects.

In this respect the Helmholtz frequency of the inlet channel can serve as a reference, although it must be mentioned that the geometry of the inlet channel is only partially comparable to a classical Helmholtz resonator. Evaluation using the Helmholtz Eq. is only a rough approach, and the dynamic properties of the sensor must also be verified by means of real measurements with the sensor.

The Helmholtz frequency is given by

$$f = \frac{c}{2 \cdot \pi} \sqrt{\frac{\pi \cdot r^2}{V \cdot (l + \pi r / 2)}} \qquad (3)$$

Here c is the speed of sound, r is the radius of the pipe, l is the length of the pipe, and V is the volume of the hollow body.

Geometrical measurements of the pressure channel yield resonance frequencies of a few kilohertz based on Eq. 3, which can be regarded as non-critical.

4.2 PPS2 Mechanical Interface

Most current commercially available pressure sensors resemble the sensor shown in Fig. 8 with regard to external geometry.

Fig. 8. Standard design and mounting template of current commercially available pressure sensors. The pressure sensor is attached to the inner wall of the vehicle door using two mounting points.

The package has two tabs, each with a hole, with which the sensor can be attached to the inner wall of the door. The fasteners are typically screws or rivets. Depending on the forces imposed by the fasteners, the holes in the flanges may be fitted with bushings to decouple the mechanical stress from the plastic.

If the sensor is mounted on the dry space side of the inner wall of the door, an external seal around the pressure inlet is usually necessary to ensure that the hole connecting the pressure inlet to the wet space is sealed. Using two fastening points helps ensure uniform, stable pressure on the seal over the vehicle lifetime.

5 Sensor module SMD187

The Sensor module contains the micromechanical pressure sensor membrane and an ASIC. The ASIC incorporates electronics for signal analysis, signal processing and a digital interface.

5.1 Construction and Sensor Module Housing

The sensor module is enclosed in a mold-premold housing (MPM housing) as shown in Fig. 9 and 14.

The MPM housing has been chosen because it is very similar to commonly used standard SOIC-housings. The footprint of the MPM housing is the same as for SOIC wide-body. The ASIC is molded into the MPM housing using methods that are comparable to the SOIC housing process. A cavity in which the pressure sensor element will later be installed is deliberately created during the molding process. The pressure sensor element and the ASIC are electrically interconnected via the lead frame, to which the required signal pins are bonded. The ASIC is bonded prior to molding and the pressure sensor afterwards.

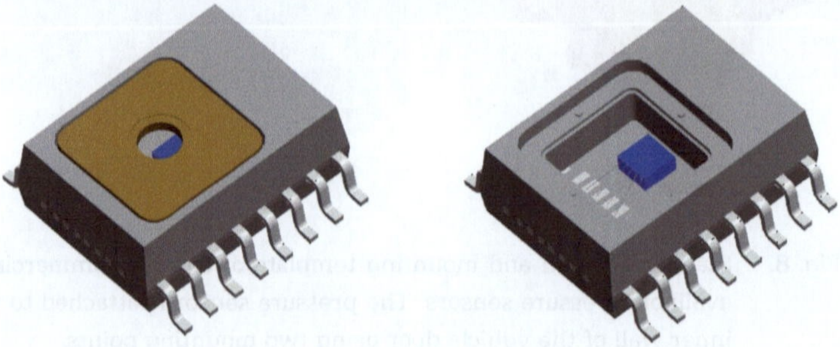

Fig. 9. Mold-premold housing (MPM housing) for the SMD187 sensor module. Shown here with and without cover.

The sensor element and the bond wires that provide the contact for the sensor element are protected from the surrounding environment by a layer of gel. The cavity containing the sensor element, bonding wires and gel layer is subsequently protected by a cover to prevent mechanical damage.

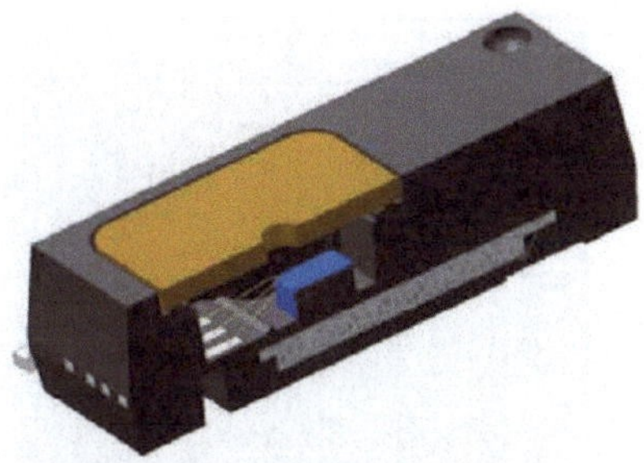

Fig. 10. Section through the SMD187 sensor module's MPM housing.

5.2 Micromechanical Sensor Element based on APSM Technology

The sensor element is manufactured from silicon using a surface micromechanical process. APSM technology (APSM = Advanced Porous Silicon Membrane) is used to manufacture the membrane. It is described in further detail in the second half of the chapter.

The sensor element is shown in Fig. 11. A pressure membrane and a temperature diode are located on the sensor element.

Pressure is measured by means of piezoresistors located on the upper side of the membrane. The mechanical strain at the edge of the membrane is thus converted into an electrical signal.

A total of four piezoresistors are mounted on the membrane. Two of the resistors exhibit a positive piezoresistive characteristic and the other two a negative one. The piezoresistors are connected in a Wheatstone bridge configuration in such a way that the different resistance characteristics of the piezoresistors are additive in regards to the output signal. The bridge is electrically energized by the ASIC, which also processes and analyzes the output signal from the bridge.

The diode is used to measure the temperature of the sensor element. The ASIC analyzes the diode output signal and the resulting temperature data is used by the ASIC to compensate temperature effects of the characteristic line of the pressure-sensing element.

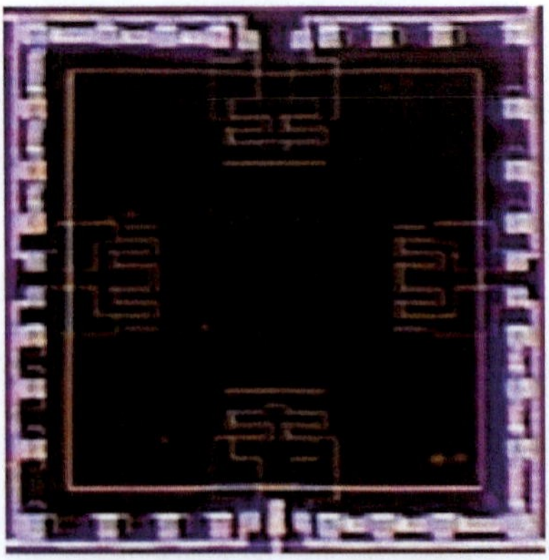

Fig. 11. Pressure membrane on the PPS2 sensor element (top view). A temperature diode is located on the sensor element in addition to the pressure membrane.

APSM technology, a surface micromechanical process, is used to manufacture the membrane.

A section through the membrane is shown in Fig. 12. The process results in the creation of a shallow cavity containing vacuum which on the top side is covered by a membrane. The membrane is approximately 14 µm thick and covers a cavity which is approximately 4 µm deep.

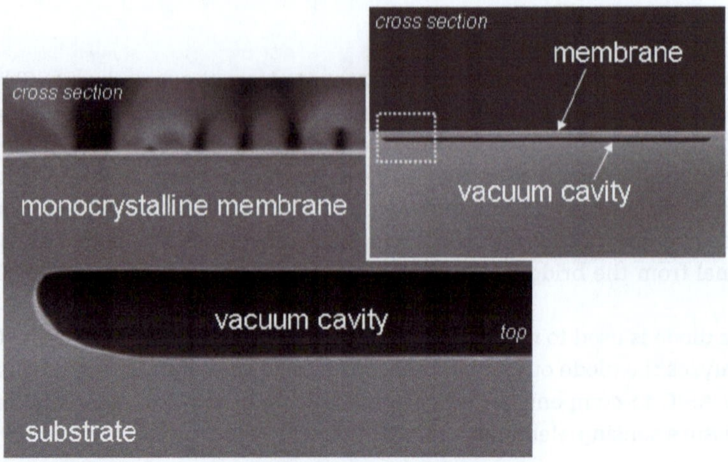

Fig. 12. Typical cross-section of the sensor membrane. It is manufactured using a surface micromechanical process (APSM = Advanced Porous Silicon Membrane).

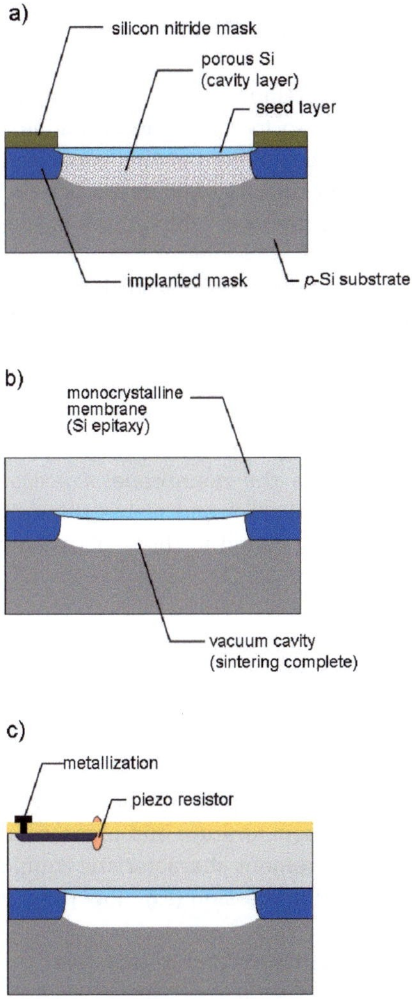

Fig. 13. APSM process, schematic representation of the steps in the membrane manufacturing process. a) after the wet chemical etching, b) after epitaxy, c) after producing the piezoresistors and metallization.

The steps required to make the membrane are shown schematically in Fig. 13.

First the membrane is structured and the monocrystalline starting layer for the subsequent membrane is formed. A wet chemical etching process is then used to etch underneath the monocrystalline starting layer creating a porous silicon layer where later the final cavity will be created.

After that, a monocrystalline silicon layer is epitaxially applied to the entire wafer and therefore also to the monocrystalline starting layer. The thermal budget for the epitaxial process and the subsequent high-temperature steps is sufficient to cause the silicon atoms in the porous silicon area to migrate and leave behind the final cavity in which there is vacuum.

The final step is to process the piezoresistors on the monocrystalline silicon layer and apply the metallization required for electrical connections and to measure the membrane deflection.

The APSM process is beneficial because it allows very robust membranes to be easily and cost effectively manufactured using purely surface micromechanical procedures. Because the membranes consist of monocrystalline silicon, they are completely leakproof and stable throughout their entire life. These properties are very well suited to the application and requirements of airbag pressure sensors.

5.3 ASIC and Signal Path

The following basic functions are integrated into the ASIC:

- internal power supply
- voltage and current supply for the sensor element
- sensor element temperature and pressure signal analysis
- pressure measurement characteristic compensation
- filter and signal processing (e.g., low pass, $\Delta P/P_0$ calculation)
- PSI5 interface
- internal diagnostics and error detection

A block diagram of the signal path in the pressure sensor ASIC is shown in Fig. 4. The ASIC used in PPS2 devices is designed so that the bridge and diode signals are digitized directly at the input and subsequent signal processing is completely digital. This results in very high design and operational reliability and is favorable because of the integration density available in modern ASIC processes.

The ASIC basically operates autonomously, as it has its own internal clock and its own power supply. It simply needs to be connected to an external voltage supply and requires an external trigger signal in synchronous mode, which initiates the data transmission.

The ASIC is produced using a high-voltage CMOS process. The ASIC and thus the sensor can be connected directly to the controller of the ECU. The externally needed circuit only consist of few passive components.

6 Technical Features and Characteristics

6.1 Operating Characteristic and Measuring Ranges

The PPS2 pressure sensor characteristic is designed for sensor operation and airbag systems used at an altitude of - 100 to 4000 m above mean sea level. The specified dynamic range for crash detection is - 5 to +15% relative to the mean ambient pressure.

These specifications result in a working pressure range of the sensor or system of 526 to 1100 mbar. The absolute pressure range covered by the sensor is equal to the absolute working range plus the dynamic range. It is 500 to 1265 mbar. The graph in Fig. 14 shows the relationship between the working range, the absolute pressure range and the dynamic range. Fig. 14 also shows that the relative dynamic range of the sensor over the ambient absolute pressure is constant. As a result, the absolute dynamic range is proportional to the ambient absolute pressure.

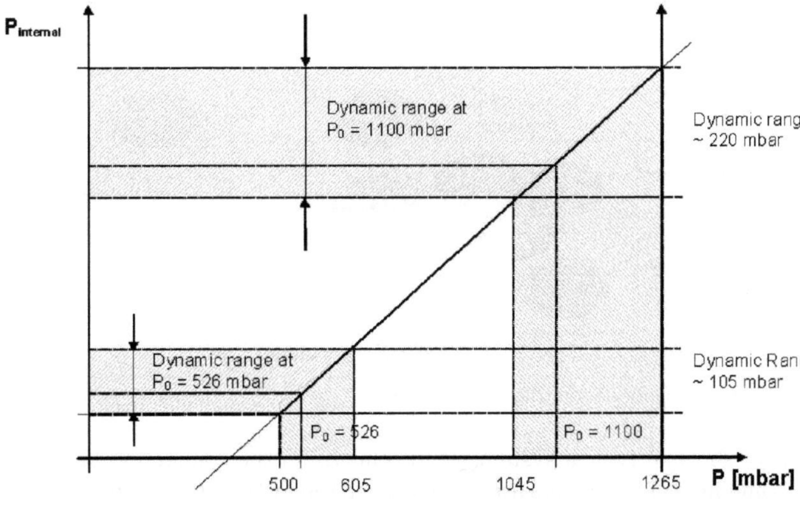

Fig. 14. Operating characteristic and working range of the PPS2 pressure sensor.

6.2 Interface, Initialization and Operating Modes

For automotive applications very high requirements apply regarding the electronic interface of sensors. Besides the data integrity the availability of the data and thus the performance of the interface plays a key role. Missing sensor data even for fractions of a second could delay the systems reaction which is not acceptable. Therefore, the data rate of the connection between sensor and ECU must be at constant level and must be safe and robust against external influences like radiated immission (EMC), ESD and others. On the other hand the cost of the interface must be as low as possible, the implementation effort should be low on sensor and on ECU-side and the interface must be flexible in regard to interchanging sensors of the same as well as of different types.

PPS2 uses a standardized current interface fulfilling all of the previously mentioned requirements: PSI5 (Peripheral Sensor Interface, 5th generation). PSI5 is an interface based on different former proprietary interfaces used in automotive applications. It is an open standard that is controlled by a steering committee and no license fees apply for using the standard. The working principle and features of PSI5 as implemented in the PPS2 are described in the following sections. Fig. 15 shows a comparison of PSI5 and other commonly used digital interface types by rating implementation cost and data transmission rate.

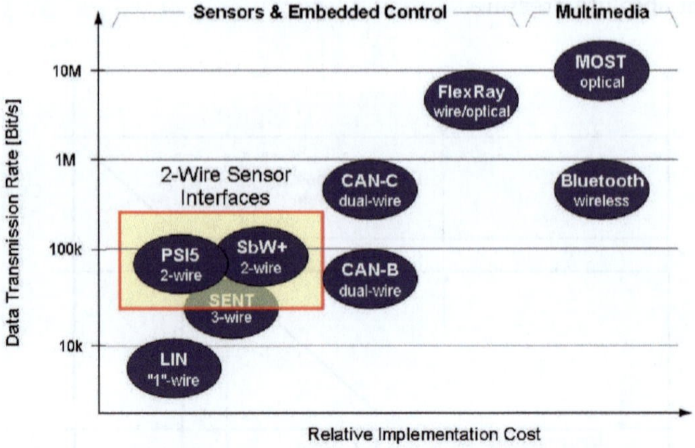

Fig. 15. Overview of digital interfaces

Fig. 16 shows a typical system setup concerning the connection between sensors and ECU. The sensors are connected to the ECU by two wires. These wires on one hand serve as power supply to the sensor. On the other hand sensor and ECU communicate over these two wires. The receiver supplies the sensors with the needed voltage and reads in the sensor data.

The example shows a point-to-point connection for sensors one and two and a bus configuration for sensors three and four which is also supported by PSI5.

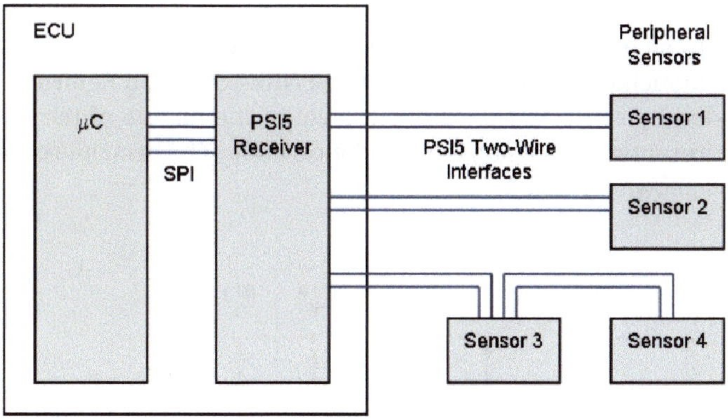

Fig. 16. Typical system setup showing examples for commonly used sensor configurations

PSI5 is based on the common principle of a current interface. The sensors are supplied by two connections to the receiver: Ground and supply line. The static condition on the line is defined by the quiescent current of the sensors connected to one receiver channel. In order to transmit data from the sensor to the ECU a current source is switched by the sensor drawing some extra current in active state. The current changes are detected by the receiver in the ECU using a shunt resistor that translates current signals into voltage signals.

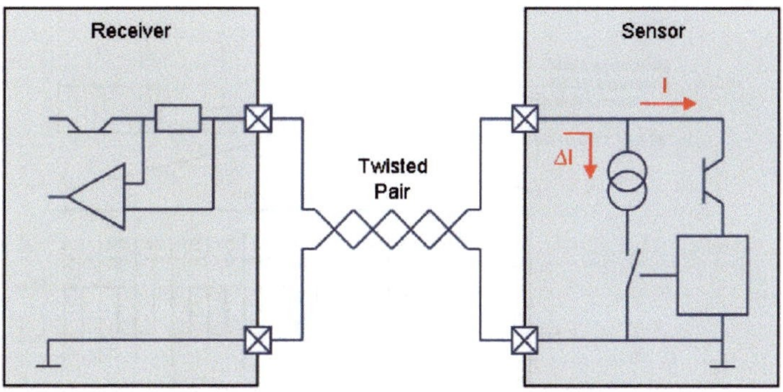

Fig. 17. Physical layer of PSI5 current interface

The relatively high sending currents of 26 mA itself result in a very good noise immunity of the interface. Only very high noise levels can disturb the data recognition and in addition high frequency noise gets significantly damped by the bus systems due to the high capacitive load in the range of 30 to 100 nF.

Using PSI5 single data bits are coded into current level switching using the so-called Manchester Code. Other than the NRZ-code (Non Return to Zero) each bit is coded by switching the current level in the middle of the bit. A logical "0" is represented by a rising edge and vice versa. The principle of Manchester Coding is shown by Fig. 18 and 23.

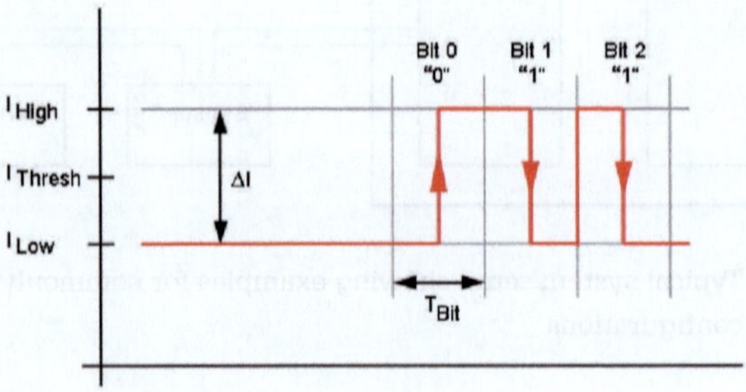

Fig. 18. Manchester Code

The data link layer of PSI5 works as follows. Each data packet consists of 2 start bits each and of an additional variable number of 8, 10, 16, 20 or 24 data bits. To ensure data integrity a parity or a 3-bit CRC-checksum are transmitted at the end of each data packet. Between different data packets a break of at least the length of one bit must be maintained. PPS2 uses solely 10 bit wide data words for data transmission as shown in Fig. 19.

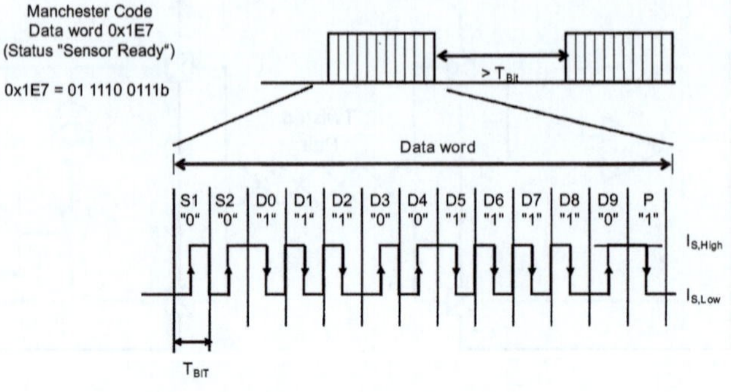

Fig. 19. 10 bit data packet in Manchester Code as used in pressure sensor PPS2

The data range definitions in PSI5 standard and specifically for PPS2 are of high significance. Safety relevant systems require a reliable distinction of sensor output data used for system algorithm and status or failure code data that is used to indicate the sensor state. The data range therefore is divided into different ranges. The data range reserved for transmitting the sensor output signal in PSI5 reaches from 480 to +480. PPS2 only uses the range of -102 to +307 which translates into the range of the normalized differential pressure of -5% to +15% relative to ambient pressure. The data ranges of PSI5 and PPS2 are described in Fig. 20.

Dec	Hex	Description	Data range
+511	0x1FF		
:	:	status- and failure messages	2
+481	0x1E1		
+480	0x1E0	highest positive data signal	
:	:	:	
+307	0x133	highest positive output signal PPS2 (15%)	
:	:	:	
0	0x000	signal amplitude „0“	1
:	:	:	
-102	0x39A	smallest (negative) pressure change PPS2	
:	:	:	
-480	0x220	smallest (negative) output signal PSI5	
-481	0x21F		
:	:	transmission of additional data status data	3
-512	0x200		

Fig. 20. PSI5-data ranges as used for PPS2 based on a data packet size of 10bits

PPS2 offers the option of transmitting the precise absolute pressure P_0 in parallel to sending the normalized differential pressure P_n. This is achieved by splitting up the 12 bit wide P_0 into 4 sub-blocks of 3 bit each, combining them with a 2 bit wide position index and coding these overall maximum 32 possible 5bit blocks into the 32 data words of data range 3 as indicated in Fig. 20.

In order not to lose any relevant P_n data the transmission of the 4 P_0 data packets gets instantaneously suppressed in case P_n reaches the internally coded noise limit.

During initialization PPS2 performs internal self-tests and transmits an electronic label to the ECU and indicates its state by appropriate status data and if needed error codes. The transmission of the electronic label is using data range 3 and includes information like sensor type code, configuration data of the sensor, manufacturer code, manufacturing date, protocol type, customer specific part code and a unique sensor serial number. In case the initialization is completed successfully the sensor switches into normal operation mode sending pressure data until it is switched off. In case an error occurs during initialization the sensor sends error messages.

PPS2 offers different operating modes described in the PSI5 standard, version 1.2. Supported operating modes are asynchronous point-to-point, synchronous point-to-point and parallel bus mode.

In asynchronous mode the sensor works independently of the ECU once switched on by the ECU when applying the supply voltage. After having performed the initialization the sensor continuously transmits pressure data to the ECU every 228 µs. The ECU receiver synchronizes with the bit-rate of the sensor by means of the two start bits.

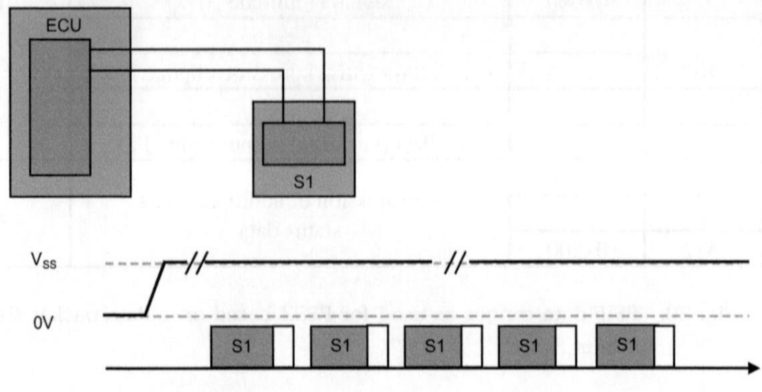

Fig. 21. Point-to-point configuration with asynchronous data transmission

In synchronous and bus mode the sensor gets periodically triggered by the ECU receiver by a voltage modulation. A sync-pulse is sent every 500 µs and between two sync-pulses 3 different time slots can be used for data transmission. The time slot to be used is determined during end of line trimming of the sensor. The advantage of the bus mode is that up to 3 sensors can be hooked up to only one receiver channel which in some cases can be very beneficial. The bus configuration set up is shown by Fig. 22.

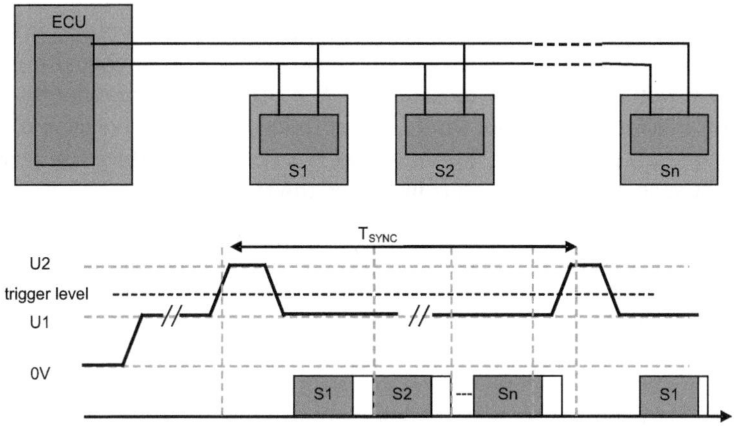

Fig. 22. parallel bus configuration using synchronous mode for data transmission

The compliance of PPS2 to PSI5 is very beneficial from a system flexibility point of view. PPS2 can be easily interchanged with for example an acceleration sensor that also complies to PSI5. If the hardware and the software are prepared to it a system can be run with either a PSI5 acceleration sensor or the pressure sensor PPS2.

7 Diagnostic Concept

The PPS2 pressure sensor performs extensive self testing during the initialization phase as well as during continuous operation. This enables the sensor to quickly detect internal sensor problems and to send appropriate error information to the controller in order to avoid erroneous system operation.

The sensor's signal path, the sensor element's measurement bridge and the temperature diode are internally tested. Furthermore the sensor conducts internal noise and signal range tests.

The transmission of absolute pressure P_0 offers a further opportunity to improve and to extend test coverage of the system.

In case the sensor transmits the absolute pressure P_0, the controller is able to compare absolute pressure data originating from the right and left doors and thus to detect errors such as drifting of the sensor's operating characteristic.

Comparing absolute pressure data of different sensors also enables the system to detect a blocked pressure inlet if the ambient pressure changes sufficiently and one of the sensors is unable to track the change in absolute pressure. A sensor would be unable to detect such a problem independently. A system using this feature will not safely detect the blocked inlet because the pressure change is a prerequisite for detection but it at least offers the possibility to detect such a failure and thus improves the test coverage.

8 Summary

Pressure sensors have been used in airbag systems to detect side crashes since 1996. They have become more and more common practice since then, particularly in midrange and high-end vehicles. Pressure sensors indirectly measure the deformation of the car body and thus are able to quickly and reliably detect side impacts. They provide a signal that is proportional to the severity of the crash and that is robust regarding interferences. Especially in cases where the point of impact is located high the pressure sensor provides a very fast signal and enables the system to quickly detect the crash and to reliably distinguish between fire and no-fire crashes.

The technical performance of Bosch's second-generation PPS2 pressure sensors for airbag applications is once again improved over that of the previous generation. The housing and construction of the sensor is based on the proven technology of the previous generation. The core of the sensor and primary source of innovation is the SMD187 sensor module, which integrates an ASIC and a micromechanical sensor element. The sensor element is manufactured using the Bosch APSM process, which is based on purely surface micromechanical process steps. The key distinguishing features of the resulting membrane are robustness and reliability. A high-voltage CMOS process is used to manufacture the new ASIC. It allows a high degree of logic integration, while at the same time requiring only simple external circuitry of few passive components.

The sensor is designed for operation of the system in heights between -100 and 4000 m above mean sea level at pressures of 526 to 1100 mbar and has a dynamic signal range used to detect crashes of -5 to +15%. The sensor performs internal signal conditioning and signal processing and transmits a normalized differential pressure signal P_n to the airbag ECU, which in case of a crash is independent of the ambient pressure.

The sensor interface is designed to use a proven and reliable 2-wire Manchester encoded current signal and is compatible with the PSI5 standard interface specification. PPS2 supports asynchronous and synchronous point to point and also parallel bus interface configuration. Due to the reduced current consumption the pressure sensor PPS2 can be exchanged with a PSI5 acceleration sensor without changing the interface hardware of the ECU.

The sensor performs extensive self testing during initialization as well as during operation. The tests are designed to detect sensor problems and prevent failures in the airbag system. The test coverage of the system can further be improved by using the option of PPS2 to transmit precise absolute pressure data P_0 and by detecting possible failures by comparing pressure data of different sensors on system level.

With PPS2 a pressure sensor for usage in airbag systems is available that has been specifically optimized for this kind of application. It provides high-speed crash data in case of side impacts and thus helps to enhance the airbag system performance while at the same time enabling easy integration into the system and the car.

References

[1] Reif, K., "Automobil-elektronik", [automotive electronics], Vieweg, second edition April 2007.

[2] J.-P. Ebersohl, C. Ohl, H. Wulff, "PSI5 – Technical Specification", Version 1.2, 06/2007, link: www.psi5.org.

[3] S. Armbruster, F. Schäfer, G. Lammel, H. Artmann, C. Schelling, H. Benzel, S. Finkbeiner, F. Lärmer, P. Ruther, O. Paul, "A novel micromachining process for the fabrication of monocrystalline Si-membranes using porous silicon", Digest Tech. Papers, Transducers 2003, June 2003, Boston, USA, 2003, 246-249.

[4] J. Marek, H.-P. Trah, Y. Suzuki, I. Yokomori, J. Hesse, J.W. Gardner, W. Göpel, "Sensors for Automotive Applications", Wiley-VCH, first edition, 2003.

[5] ISO/IEC, "Kommunikation offener Systeme - Basis-Referenzmodell", ISO/IEC 7498-1, 1994-11.

[6] SAE, "SENT - Single Edge Nibble Transmission for Automotive Applications", SAE J2716, 4/10/2007.

[7] ISO, "Zünd- und Sensorbus für Insassenschutzsysteme", ISO 22896, 1994-11.

[8] Autoliv, Bosch, Continental, Siemens VDO, "PSI5 Technical Specification V1.2", 14.06.2007, www.psi5.org.

[9] VDE, "Funk-Entstörung zum Schutz von Empfängern in Fahrzeugen, Booten und Geräten - Grenzwerte und Messverfahren (IEC/CISPR 25:2002)", VDE0879-2, 2003-11.

Boris Adam, Tobby Brandt, Ralf Henn
Sebastian Reiss, Markus Lang, Christian Ohl
Robert Bosch GmbH
P.O. Box 1342
72703 Reutlingen
Germany
Boris.Adam@de.bosch.com
Tobby.Brandt@de.bosch.com
Ralf.Henn@de.bosch.com
Sebastian.Reiss@de.bosch.com
Markus.Lang@de.bosch.com
Christian.Ohl@de.bosch.com

Keywords: airbag pressure sensor, PPS2, micromechanical pressure sensor, advanced porous silicon membrane, surface micromechanical process, microcrystalline membrane, integrated signal conditioning and self test, PS15 standard interface, sensor module SMD187, mold-premold housing SOIC16

Trusting Your Senses

G. Hardman, Silicon Sensing Systems Limited

Abstract

The concept of signal integrity as a measure of trustworthiness of data is described using a simple analogy. Contrasts are drawn between integrity and accuracy in regard to the use of the data in a safety related application. The concept of "false plausibility" of data is described, with associated failure modes of the analogy examined. An implementation of a yaw rate sensor is described that takes advantage of the use of continuous numeric analysis of the control loop variables to provide very high levels of loop observability in real time. It also considers detectability of failures in times significantly less than the time constant of the sensing system, i.e. fault detection that is not bandwidth limited. Also considered are feedback as a means of verifying the conversion processes within the sensor, and parallel verification of the operation of the sensor software and microcode by using pseudo-random seed data from the sensor.

1 Introduction

In everyday situations, there are often times when an action or reaction relies critically on the ability to sense both accurately and reliably. Decisions that affect safety are based on the trustworthiness of the data. It isn't acceptable to have a suspicion that something is true, it is essential to be sure beyond a basic level of trust.

Take an occurrence that happens millions of times each day. A driver wishing to join a flow of traffic on a major road must confidently and reliably know that it is safe to do so. At junctions fitted with traffic lights, one level of trust is that when the lights are green to join the major road, oncoming traffic is inhibited by a red light. However, it cannot be guaranteed that it is safe to join, as it depends on all other drivers obeying the rules: stop on red. The detection of a green light gives no data as to the safety of joining, as it doesn't have any information on the rest of the traffic environment. Emergency vehicles may legitimately continue against a red light; an inattentive driver might do so regardless.

So, what is needed is reliable, trustworthy data. Not necessarily accurate data; there is no need to estimate to within a couple of metres where an oncoming motorcycle is, only whether it is sufficiently far away and travelling sufficiently slowly that it is safe to pull out in front.

What this sensing scenario requires can be dubbed "integrity". That is, a dependence on having data that is sufficiently accurate for the purpose, but with a quality aspect that means that the data can be used unambiguously to initiate or control an event. A high integrity sensor is one which gives an adequate level of performance in terms of the normal parameters, (sensitivity, offset, noise) but in addition provides unambiguous data. A low integrity sensor might be more accurate, (more sensitive, lower noise), but cannot be trustworthy if there are any circumstances where the output is "false plausible".

Data	*Failure mode*	*Plausibility check*
Can I see anything coming?	View obscured, (window fogged, parked truck).	Can I see anything at an appropriate range?
Can I see anything moving towards me?	One glance giving time invariant data.	Look again, can I see movement? Has the scene changed?
Can I predict that the gap is wide enough to give time to pull out?	Wet road, lost traction, bad estimate of own acceleration.	Historic data trend (i.e. change in range of oncoming vehicle, responsiveness at previous junction) correlation with other data, (wipers on icy).
Are there other characteristics of the data that should inhibit pulling out?	Oncoming emergency vehicle signalling to turn.	Signal characteristic analysis, (flashing blue lights, turn signals, vehicle road positioning).

Tab. 1. Failure Modes in joining traffic

The concept of "false plausible" is an essential factor of any system that has safety implications. In particular, it must be considered in the system's failure modes as well as its normal operation.

Let's return to the junction. Consider a traffic light system employing only red lights; here a "no red light" condition may be interpreted as permission to proceed. This is fine in normal operation, but has an obvious "false plausible" failure mode – the bulb fails or there is no power. Now, "no red light" only means that it cannot be said that it is definitely unsafe to proceed, as the data is ambiguous. Even having a green light still has "false plausible" failure modes. Whilst a green light shows that the lights are active, (and even assuming that the designers have a fail-safe that prevents the lights from being green both ways), there are still the failure modes of the inattentive driver or the oncoming emergency vehicle.

The requirement for this environment is therefore a sensor having "integrity". This is a combination of design parameters, methodology and an intimate understanding of the failure modes. Some of the failure modes are given in Table 1. For the driver in our scenario, the primary sensor is the eye, (in partnership with the cognitive functions of the brain). Now the signal from the eye can be checked for plausibility.

So having a very accurate sensor, such as excellent eyesight, is insufficient if there are inadequate checks and cross checks. A skilled driver adds integrity to the observed scene.

2 Redundancy – The Limited Option

Some systems attempt to improve the integrity of the data by adding redundancy to the sensing. However, the use of two similar sensors can address only a small proportion of the failure modes, leaving many of them unchanged. This is because the use of two similar sensors cannot overcome "common cause" failure mechanisms. Consequently, there is often little improvement in integrity for the system as a whole. In the example above, a driver having two eyes has much greater defence against poor range estimates, as the brain performs a ranging function on the binocular image. This addresses some of the third set of failure modes, as movement towards the observer is easier to detect, but does little or nothing for the other three sets. Unless the view is obscured by a very narrow object, having two eyes close together doesn't add much observability to the scene. Redundant systems have a defence against total, detectable signal loss, but have no added defence against false plausible failures.

Following the weight of evidence regarding the improvement in safety, Electronic Stability Control (ESC) is rapidly gaining prevalence in braking systems. This has the ability to make certain accident types entirely avoidable, and is being mandated by NHTSA for fitment into passenger cars in the US, with other markets introducing such systems as standard. Clearly, in order to improve the safety of a vehicle, the ESC system must be reliable and trustworthy. It must have high integrity.

Reduced to its simplest, an ESC system comprises the means of modifying the braking forces at each wheel to reduce understeer or oversteer if the actual behaviour of the vehicle varies from that desired by the driver. The primary mechanism for detecting this variation is by inertially sensing the actual dynamics of the vehicle, (lateral acceleration and yaw rate) as distinct to that commanded by the driver, (primarily steering wheel angle) and the response at the wheels, (wheel speed). The absolute minimum requirement for an ESC system is the ability to unambiguously measure the vehicle yaw rate to validate the response to steering inputs.

In this application, it is essential that any failure mode of the sensor results in an error that is detectable or that results in a signal that cannot be regarded as "false plausible". If the sensor fails completely, i.e. fails "silent", then the system has the potential to avoid making erroneous braking demands. The system does not improve the safety of the vehicle in this instance, but neither does it make it worse. The worst case scenario is if a failure mode exists that causes a false plausible yaw rate output, which provokes the system into making braking demands that are not required. In this case, the vehicle may be forced into over or understeer without any warning or input from the driver.

3 High Integrity Rate Sensing

Silicon Sensing's VSG4 family of sensors has been developed specifically to address this issue. Throughout the design and development, key to the process has been the identification and mitigation of potential error sources and false plausible failure modes. Several techniques are employed that have given rise to a high integrity sensor, or the ability to provide high integrity data when integrated into a system. An overview of the operation of the yaw rate sensor helps in understanding the techniques to improve integrity.

The sensor uses a bulk silicon micro-machined ring structure and a mixed signal ASIC together with a signal conditioning processor to make a single-axis yaw rate sensor.

3.1 Principle

The silicon ring is driven in a cos 2Θ mode shape to produce the radial velocity components required to make a Coriolis gyroscope.

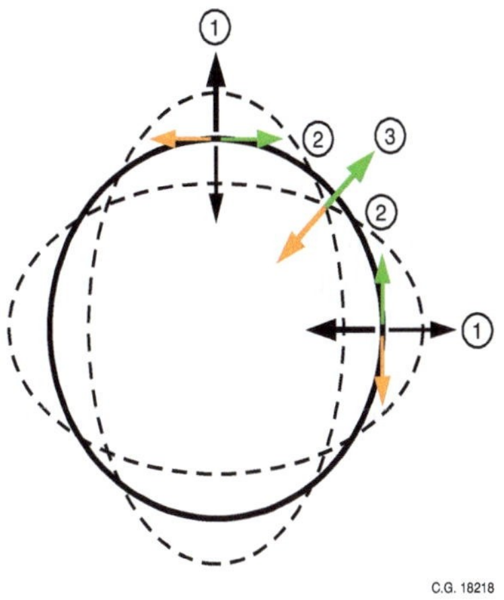

Fig. 1. Vibration mode

Fig. 1 shows the movement of the silicon ring while vibrating:– As the ring oscillates in its natural state, the ring moves in the way shown by (1). When the gyro is rotated about its sense axis, the Coriolis force acts tangentially on the ring as shown by (2). The vector sum of these forces result in a combined resultant vibration 45° out of alignment with the primary vibration, as shown by (3).The force required to null this resultant vibration completely is directly proportional to the angular rate.

3.2 Implementation

Eight uniformly spaced transducers are placed around the sensor ring structure to form two pairs of drive transducers and two pairs of pick-off transducers. VSG4 uses electrostatic transducers, where the forces are detected or created by charged capacitor plates.

One pair of diametrically opposed drive transducers is used to excite the cos 2Θ mode with the phase and amplitude sensed by the corresponding pick-off transducer pair. The drive amplitude and frequency is controlled by the electronic control loop to establish an accurate radial velocity component at the resonant frequency of the structure. Frequency is set by a Phase Locked Loop (PLL), operating with a Voltage Controlled Oscillator (VCO), whilst amplitude is controlled using an Automatic Gain Control, (AGC). The primary loops set the basic operating point and scale factor of the system.

Unlike more basic sensors that attempt to measure the rate of turn directly, and are unable to eliminate internal errors, all Silicon Sensing gyros use a servo to measure the force required to completely cancel out the secondary vibration. To do this, a secondary rate nulling loop is set up using the other drive and pickoff transducer pairs to enable the Coriolis forces (generated by applied angular rate) to be detected and servoed to zero using digital filters within the processor. This fully closed loop operation largely removes dependency of performance on the mechanical amplification factor, Q, of the resonator and enables excellent bias and scale factor performance to be achieved without any compensation. The force required to null the Coriolis force is a very accurate measure of the applied angular rate. From an integrity viewpoint, null is the least ambiguous state to determine!

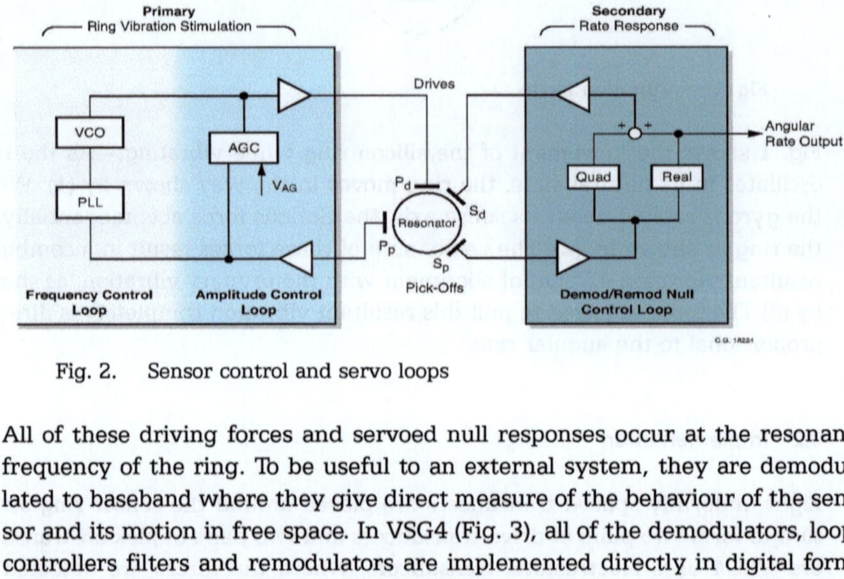

Fig. 2. Sensor control and servo loops

All of these driving forces and servoed null responses occur at the resonant frequency of the ring. To be useful to an external system, they are demodulated to baseband where they give direct measure of the behaviour of the sensor and its motion in free space. In VSG4 (Fig. 3), all of the demodulators, loop controllers filters and remodulators are implemented directly in digital form within an internal digital signal processor, (DSP). The yaw rate and telemetry regarding the status of the control loops are available directly in numerical form. This more sophisticated technique eliminates performance variation or drift typical of analogue control loops, as the numeric control is inherently free from ageing, temperature and manufacturing tolerances. All numeric processing is synchronised to the MEMS resonant frequency, which eliminates the temperature-sensitive, phase-related detector errors commonly found in other MEMS sensors.

The derived rate data is digitally filtered, scaled and then output via the Serial Peripheral Interface (SPI), bus . For legacy applications, the rate data is also converted to analogue format via the Digital-Analogue Converter, (DAC) (with factory programmable scale factor and bias) and output at separate pins.

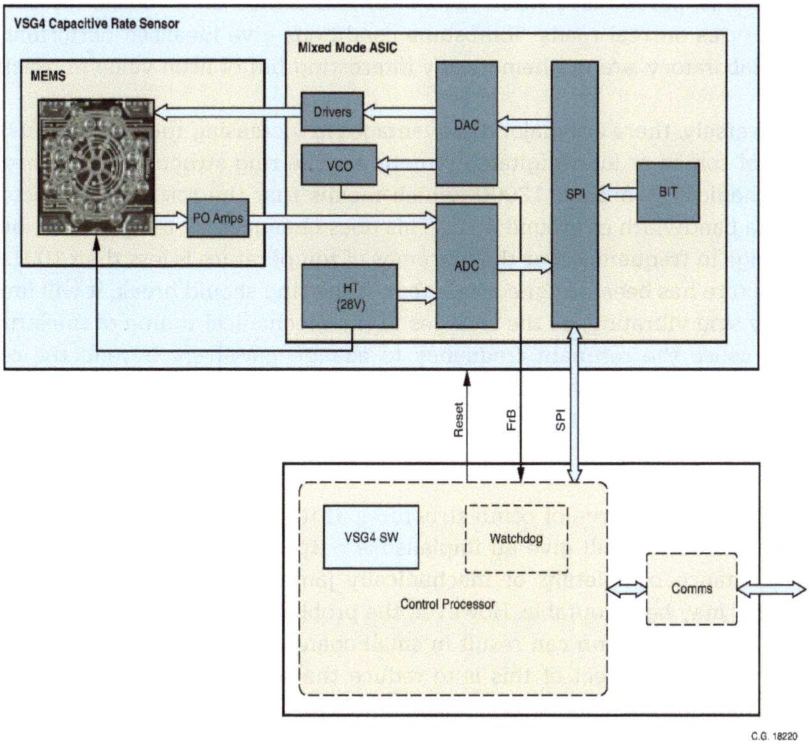

Fig. 3. VSG4 system block diagram

3.3 Structured Integrity

As the sensor system is inherently digital, it lends itself to several independent techniques to ensure that false plausibility is minimised. These operate continuously, without interrupting or otherwise corrupting the rate output. The nature of the ring structure itself is such that failure mechanisms are catastrophic rather than parametric in nature, as can be the case with forks or combs, again eliminating the possibility of false plausible responses.

The ring structure has a single moving surface mounted on eight compliant spokes. This has the disadvantage that the signal levels are low, inherently due to the small capacitances involved. In artificial conditions, such as in the laboratory, this results in comparative tests that suggest that such sensors have high levels of noise. This is true in so-called "tombstone" conditions, where the sensor is immovably fixed to a massive stationary object, (the "tombstone"). This reduction in signal to noise ratio is largely irrelevant in real world environments, particularly when the sensor is mounted on a vehicle travelling on real tyres on real roads. Tombstone conditions give idealised performance in the laboratory, are mathematically interesting but of little value in practice.

Conversely, there are major disadvantages in increasing the signal level by the use of combs or inter-digitated structures. The ring structure has a very high mechanical Q, around 12000, which means that the resonant characteristic has a bandwidth of around 1 Hz. This does change with temperature, but the change in frequency over the extremes of temperature is less than 10 Hz. The structure has been designed such that, if the ring should break, it will immediately stop vibrating, as the changes in the mechanical tuning of the structure will cause the resonant frequency to change massively beyond the control range of the primary phase locked loop. Also, as there is only one interface to the structure, it isn't possible to lose part of it and continue to operate with reduced sensitivity. It will fail completely.

The same is not true of comb structures. If it can be guaranteed that a failed comb structure will give an implausible output, such as by shorting out the capacitance completely, or mechanically jamming the motion, then such a sensor may be acceptable. However, the problem is that failed or disconnected teeth within a comb can result in small changes to the overall capacitance of the plate. The effect of this is to reduce the sensitivity of the sensor whilst not inhibiting its output. This will inevitably give rise to false plausible output that cannot be detected without some other inertial reference, i.e. a second dissimilar sensor.

3.4 System Node Observation

Implementing the servo control in the digital domain enables every "node" of the system to be observable. Analogue control loops can only be examined internally for plausibility by the use of additional comparators. In that way, it is possible to determine when the control loop has completely saturated, but only by the addition of more circuitry, all with its own failure rate and error budgets. Such techniques are costly and risk unacceptable false-alarm rates. Whereas, by having every term expressed as a variable, every node can be checked for reasonableness, plausibility and time-variance. In this way, a term that has become stuck is drifting, or has simply gone out of specification, can be detected and reported via the SPI interface. As all of this can be achieved within the DSP code, there is no additional circuitry required, no additional time delay and much more opportunity for unambiguous fault detection. Furthermore, these reports form part of the standard payload of the SPI payload, and are reported within every message, at one millisecond intervals. Fig. 4. shows how this can be achieved for a full four-loop control system.

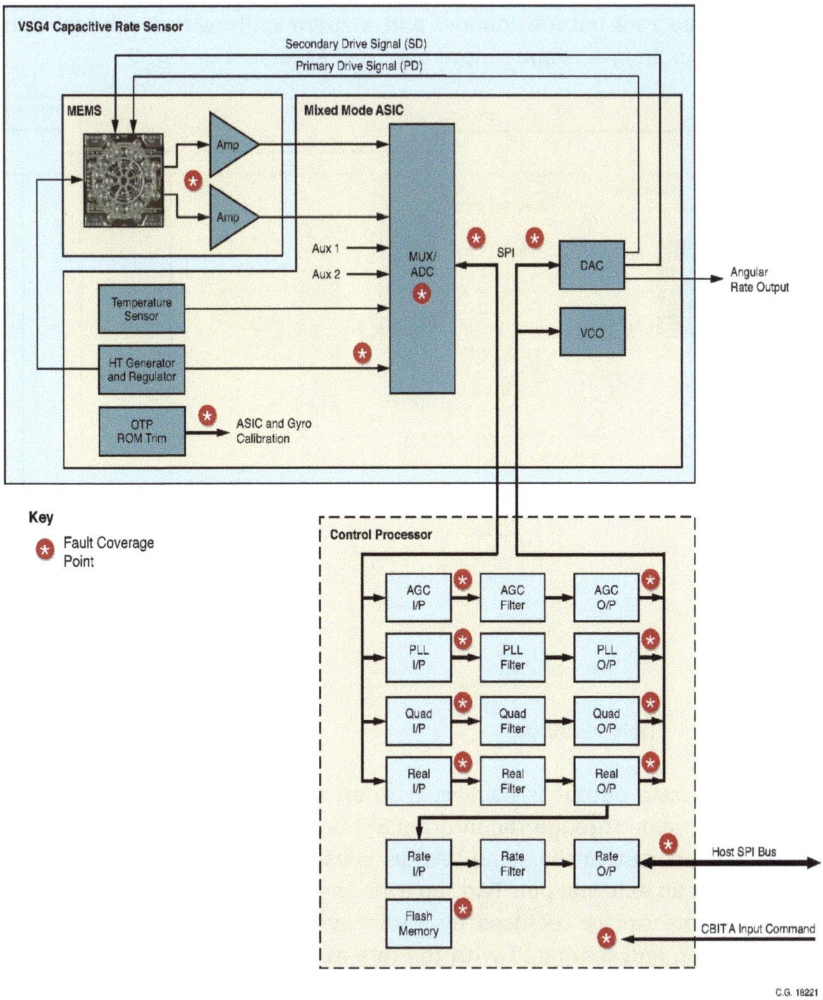

Fig. 4. Observable nodes

3.5 Analogue Feedback

The sensor block diagram shows that the voltages from the transducers (which contain the primary motion and Coriolis force information) feed through an ADC before being passed to the DSP. The return signals, (i.e. the feedback voltage demands for the servo loops) are converted to an analogue voltage by a DAC. This signal chain, together with that portion of the DSP software associated with conversion and internal communication, can be tested by feeding the rate signal back through the same signal path. To do this, use is made of the analogue rate output channel and a spare multiplexed channel into the ADC, the auxiliary analogue input channel, as shown in Fig. 5.

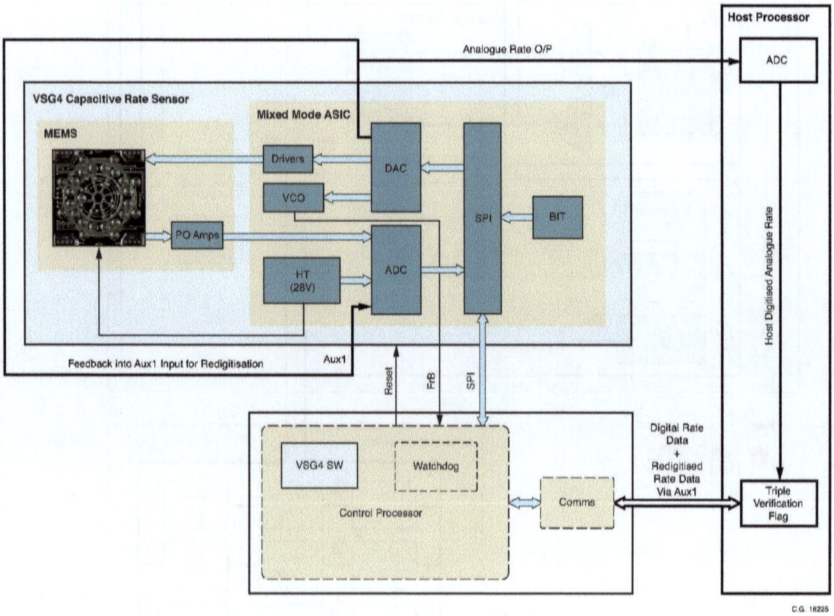

Fig. 5. Triple verification

The digital rate output is converted to an analogue voltage by passing the digital rate value through the internal SPI bus from the DSP to the ASIC. This value is converted in the same DAC as is used for force-rebalance, and made available at an external pin. Two separate checks are now possible. Firstly, the analogue rate can be digitised by a host system, such as the ESC Electronic Control Unit, and compared with the rate as reported over the SPI bus. There is a little signal degradation in the conversion process, but the result will be sufficiently close to confirm that the output of the conversion from digital to analogue is plausible. Secondly, this analogue signal can be connected to an auxiliary input of the same ADC as is used to monitor the transducers. It is redigitised within the sensor and presented to the host as an auxiliary data field in the SPI payload. Consequently, there are three measures of yaw rate that can be compared. If the redigitised rate disagrees with the digital rate, then the sensor operation is implausible. If the host's digitised rate disagrees with the redigitised rate, then the host operation is erroneous.

This technique also checks many of the op codes for the DSP, since corruption of any of the communication, loading, register manipulation or scaling software will result in a corrupted analogue output, which is consequently detectable.

3.6 DSP Verification

The control loops for the sensor are implemented as executable code within a very small DSP, running synchronously to the sensor resonant frequency. The clocks for the DSP are derived from the sensor's primary resonance, (being multiplied up 1024 times by phase-locked loop). Complete catastrophic failure of the DSP is very easy to detect, as this would be a "fail silent" situation. Parametric errors in the variables within the servo loop are detected as described above. There remains the failure mode whereby the DSP could perform an operation incorrectly, for example corrupt an operation of the arithmetic unit, such that the result of a calculation is in error. In this instance, without mitigation, the sensor would report an incorrect rate output with no means of internally validating the result.

This is overcome using a technique that is employed by every child on encountering a calculator for the first time. Children use the calculator to multiply 2x2 and smile when they get the answer 4. They use known input data to give an expected result, verifying the input, manipulation and output of the data. However, this is an example of a test that can give a false plausible output. Consider the case where the operator may have failed, (x becomes +), but the answer is correct, (still 4). The next thing they try is 3x3. This is to ensure that the system can respond to a VARYING input and give a different, but still predictable, result. Better still, this calculation is less prone to ambiguity. Hence by checking that 3x3 isn't 6, the operator has been verified. This is essential in tests for high integrity, as "stuck" data can be hard to detect – especially if it is plausible.

Obviously, with only one arithmetic unit, the sensor cannot check for this in isolation. However, the sensor has a ready source of time variant data that can give a predictable result to a test calculation. A scheme for using this time-variance is shown in Fig. 6.

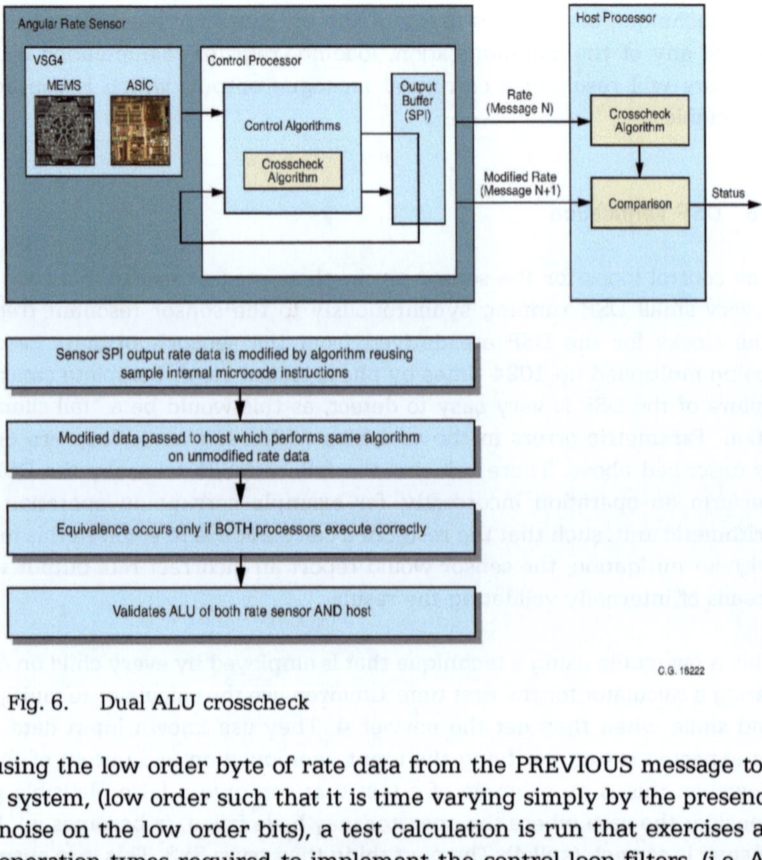

Fig. 6. Dual ALU crosscheck

By using the low order byte of rate data from the PREVIOUS message to the host system, (low order such that it is time varying simply by the presence of the noise on the low order bits), a test calculation is run that exercises all of the operation types required to implement the control loop filters, (i.e. ADD, SHIFT, MULTIPLY, LOAD), and a packet of data can be generated that is completely predictable by the host, simply by implementing the same calculation. The output from this test calculation is included in the message payload to the host. In the mean time, the host will have received the seed data in the previous payload, run the same algorithm, and can compare the result at the next message cycle. The comparison is a simple equivalence function. For the test to pass, the calculation in both the host and the sensor must have been performed correctly or must have failed in exactly the same manner. As the host will not have an identical tiny DSP, the probability of a false plausible result is extremely low as there are no common cause failure mechanisms. In this way, whilst it is not possible to isolate a failure to the host or the sensor ALU, it is very practical to verify that the data transmitted is plausible.

3.7 Commanded BIT

The failure identification mechanisms described in the preceding sections operate continuously without interrupting or modifying the rate output. This is regarded as Continuous or Background Built In Test (BIT). There is also a commanded BIT, which is provided for backwards compatibility with legacy systems. Its operation mirrors many of the commanded BIT implementations in other sensors, but doesn't offer any advantages on its own.

In response to an external command, (either via the SPI interface or in response to a logic input), the secondary loop is disturbed to artificially invoke a fixed rate offset. This causes the output to change by a predetermined offset, which can be detected by the host system either via the SPI interface or via the analogue output. However, on its own in any system, Commanded BIT is a poor indicator of output plausibility. It has value in determining that a sensor is basically operational as part of a power up sequence. It is well known that many failure modes in electronic systems are the result of power cycle stresses, so Commanded BIT can detect a sensor that failed catastrophically at power-up or previous power-down.

In this regard, it is rather like an annual medical exam; it has a value in terms of an assessment of operation at one off-line instant in time, but gives no information regarding the ongoing health of the patient, nor whether they are fit to undertake some activity. Many sensors have a commanded BIT that causes the sense element to respond so as to give a maximum output, (i.e. the analogue output goes to rail). These have value only in that they confirm that the sensor is wired up.

3.8 Message Structure

Using the techniques described above, the sensor collates and exports a data payload to the host. The final link in the chain is to ensure that the data payload is also free from "false plausible" errors.

As the data payload includes status, rate output, the test calculation modified rate, the reconverted analogue rate and the usual communication niceties of checksum and command handshake, there remains one final error mechanism. If the sensor continuously outputs a stable, invariant data packet without corruption, then the only way to trap this is to compare the data with some historic trend. Curiously, this would result in a very stable gyro giving false alarms, as the data might not change perceptibly, particularly if the vehicle was stationary at the time.

This is easily overcome by inserting a simple message counter (even if only two bits), such that the message cannot avoid having a time variant component. Not only would this ensure that the checksum is exercised, but it allows the host to simply compare adjacent messages for evidence of update. By adding the counter into the host checking, the input buffer of the Host can also be verified as not having stuck data.

4 Summary and Conclusion

All of the aspects considered in the "traffic light" scenario can be applied to a yaw rate sensor designed with high integrity from the outset. Simply replicating sensors to add redundancy does not result in significant improvements in integrity. False plausibility has been categorised, and mitigating operations included for each category. The rate signal can be checked to be in a realistic range, derived from information that has itself been checked against control limits. The manipulation of the data to form a measurement can be checked by performing a parallel verification calculation, and simple steps can be taken to ensure that it isn't stuck. Finally, the data can be "peer reviewed" by sharing variables with the host and manipulating them in a low-overhead manner. Far from being an academic exercise, these techniques have been successfully embodied in Silicon Sensing's VSG4 production for cost effectively providing yaw sensing with high integrity.

Geoff Hardman
Silicon Sensing Systems Limited
Clittaford Road, Southway
Plymouth PL6 6DE
Devon
Great Britain
geoff.hardman@siliconsensing.com

Keywords: MEMS, silicon, sensor, gyro, ESC, integrity, safety, redundancy, processor, failure mode

Appendix A

List of Contributors

List of Contributors

Appendix B

List of Keywords

List of Keywords

MIX
Papier aus verantwortungsvollen Quellen
Paper from responsible sources
FSC® C105338

If you have any concerns about our products,
you can contact us on
ProductSafety@springernature.com

In case Publisher is established outside the EU,
the EU authorized representative is:
Springer Nature Customer Service Center GmbH
Europaplatz 3, 69115 Heidelberg, Germany

Printed by Libri Plureos GmbH
in Hamburg, Germany